Lecture Notes in Mathematics 1596

Editors:
A. Dold, Heidelberg
B. Eckmann, Zürich
F. Takens, Groningen

Lutz Heindorf   Leonid B. Shapiro

Nearly Projective Boolean Algebras

With an Appendix
by Sakaé Fuchino

Springer-Verlag

Berlin Heidelberg New York
London Paris Tokyo
Hong Kong Barcelona
Budapest

Authors

Lutz Heindorf
Sakaé Fuchino
2. Mathematisches Institut
Freie Universität Berlin
Arnimallee 3
D-14195 Berlin, Germany

Leonid B. Shapiro
Department of Mathematics
Academy of Labour and Social Relations
Ul. Lobachevskogo 90
117454 Moscow, Russia

Mathematics Subject Classification (1991): Primary: 06E05

Secondary: 54A35, 54B20, 54C55, 54D35

ISBN 3-540-58787-X Springer-Verlag Berlin Heidelberg New York

CIP-Data applied for

This work is subject to copyright. All rights are reserved, whether the whole or part of the material is concerned, specifically the rights of translation, reprinting, re-use of illustrations, recitation, broadcasting, reproduction on microfilms or in any other way, and storage in data banks. Duplication of this publication or parts thereof is permitted only under the provisions of the German Copyright Law of September 9, 1965, in its current version, and permission for use must always be obtained from Springer-Verlag. Violations are liable for prosecution under the German Copyright Law.

© Springer-Verlag Berlin Heidelberg 1994
Printed in Germany

Typesetting: Camera-ready $T_{E}X$ output by the authors
SPIN: 10130239   46/3140-543210 - Printed on acid-free paper

To our parents

Preface

The history of this publication starts in April 1992, when the second author gave a talk at the Freie Universität Berlin.

As one result of this lecture part of the audience felt the desire to study some topological papers in more detail than before. A series of seminar talks was given by the first author and the present text grew from the accompanying notes.

The idea to turn those notes into a joint publication occured after the second author obtained some new results that would nicely complement the already gathered material. Work on this project was then begun separately in Moscow and Berlin with interchanges by mail.

The actual writing of the text was done by the first author who has to thank several people for their help. Sabine Koppelberg and Sakaé Fuchino read previous versions and gave helpful comments, which led to simplifications in some proofs. Many valuable remarks also came from Ingo Bandlow.

John Wilson improved the English and Ulrich Fuchs helped with TEX.

The final version was prepared jointly by both authors during March 1994 in Berlin. We want to thank the Deutsche Forschungsgemeinschaft for its financial support (Grant Number 436 RUS 17/192/93), which made the visit of the second author possible.

After the main text was ready, Sakaé Fuchino kindly wrote an appendix on set-theoretic methods in the field, which also includes some of his recent independence results.

Berlin, September 21, 1994   L. H.   L.B.S.

Contents

Introduction

The first manifestation of interest in projective Boolean algebras seems to be the paper [22] by Halmos. He established the (now) familiar properties of projective objects in general and proved that all countable Boolean algebras are projective. This is in sharp contrast to what happens for other classes of algebras, where difficult questions arise already at the finite level.

The decisive tools for the study of projective Boolean algebras came from topology. As projective Boolean algebras can be embedded into free algebras, their dual spaces are dyadic. This class of spaces had been introduced by P. S. Alexandorff and studied by topologists for many years. Most results about them concerned cardinal functions in one form or the other. For a satisfactory structure theory the class of dyadic spaces turned out to be too wide, however. For the subclass of AE(0) spaces (its zero-dimensional members are exactly the dual spaces of projective Boolean algebras) such a theory started with the remarkable paper [23] by R. Haydon. He established that these spaces admit a special kind of inverse limit representation. By means of this description Haydon showed that the class AE(0) coincides with the class of so-called Dugundji spaces, which were earlier introduced by A. Pełczynski ([44]) via a functional analytic property. The name 'Dugundji space' has later become popular and is in the given context often used instead of AE(0).

Haydon's inverse limits $X = \varprojlim \{X_\alpha; p_\alpha^\beta; \alpha < \beta < \lambda\}$ have an ordinal as index set and are continuous, i.e. for limit ordinals $\gamma < \lambda$ the space X_γ is homeomorphic to the limit of the restricted system $\{X_\alpha; p_\alpha^\beta; \alpha < \beta < \gamma\}$. Such inverse systems are called 'transfinite spectra'. The most important feature of Haydon's spectra is, however, the specific nature of the bonding maps $p_\alpha^{\alpha+1}$.

Transfinite spectra were probably first used by L. S. Pontryagin under the name 'Lie series' in his analysis of the structure of compact groups. Before Haydon's paper transfinite spectra were used by S. Sirota to characterize the Cantor cube of weight $\aleph_1$. As a consequence of his characterization Sirota proved that the hyperspace (or exponential) of a dyadic space of weight at most $\aleph_1$ is dyadic again. This result, and the obvious question of what happens for bigger weights, had an essential influence on the further development of the theory.

In the mid seventies, investigating uncountable products of metrizable spaces, E. V. Ščepin introduced the class of κ-metrizable spaces. Using Haydon's characterization he proved the κ-metrizability of all Dugundji spaces. This led him to a spectral characterization of compact κ-metrizable spaces and to general questions about inverse limit representations.

Generalizing the concrete work of his predecessors Ščepin introduced the concept of a class of compact spaces and a class of continuous mappings being 'adequate'. Roughly speaking, spaces are classified according to whether they admit inverse limit representations in which the bonding maps are taken from a special class of mappings. This idea, in some sense, reduces the study of spaces to the study of mappings.

The class of Dugundji spaces is adequate to what Ščepin called 0-soft mappings. We shall be mainly concerned with the adequate pair that consists of κ-metrizable (otherwise known as open generated) spaces and open mappings. It was studied mainly by Ščepin, with important contributions coming from other Moscow topologists, L. V. Širokov and A. V. Ivanov to name just two. The latter established an important link between κ-metrizable and Dugundji spaces: a compact space is κ-metrizable iff its superextension is a Dugundji space. In fact, this theorem was an important step in the proof of the adequateness.

Another adequate pair, which will play a prominent role below, grew out of the theory of absolutes and co-absoluteness, which dates back to I. V. Ponomarev [45]. It turned out that the class of spaces co-absolute to Dugundji spaces is adequate for the class of mappings co-absolute to 0-soft ones. The starting point here was the second author's result that every dyadic space is co-absolute with (any compactification of) an at most countable sum of Cantor cubes of suitable weights.

In this work we shall use the language of Boolean algebras to present most of the results and concepts mentioned above. In other words, we confine ourselves to the zero-dimensional case. Due to that special case, many proofs become technically more transparent, which makes the ideas come out more clearly. Admittedly, some of the ideas that are important for higher-dimensional spaces get lost. Obviously, we do not touch the geometrical role of Dugundji spaces, more precisely their subclasses AE(n). For information about these aspects the interested reader may consult Dranišnikov's paper [10].

To assure the topologist reader that he is not wasting his time, it should be added that many interesting examples and counter-examples of the theory are zero-dimensional anyway.

We have tried to collect all the results that have been obtained over the years, mostly by Russian topologists. A word of caution is in place here. We often attribute Boolean algebraic results to topologists. This is to be understood in the wider sense. In most cases the corresponding topological result is more general being true for spaces of arbitrary dimension. In some cases, the corresponding topological result is only near in spirit to what we do.

The algebraic dual of an inverse limit decomposition of a space is the representation of an algebra as the union of a system of subalgebras. The basic idea of dealing with uncountable algebras by looking at suitable systems of well-embedded subalgebras has, independently of topological considerations, been developed by model theorists (cf. [38]) and set-theoretically oriented algebraists (cf. [11]).

One of the popular areas in this field is the study of almost free algebras in various classes. This is in spirit similar to what we do in the text, but will play no explicit role. The reader may consult [18] for further information and results concerning these problems for the class of Boolean algebras.

Having returned to the algebraic setting we have to mention the name of S. Koppelberg, who solved some of the problems from [22] and cultivated the technique of decomposing Boolean algebras into chains of subalgebras. Her survey [35] in the *Handbook of Boolean Algebras* made much of the material about projectivity available to people working in algebra and set-theory. The present work, which mainly deals with generalizations of projectivity, can be regarded as a continuation in the same direction.

We now briefly introduce and motivate the main concepts that will play a role in this work. We also try to summarize the contents in order to give the reader an idea of what he can expect. Some of the statements in this introduction may contain unexplained notions or be otherwise somewhat imprecise. Everything really needed will be vigorously repeated in the main text.

Overview

Projective and rc-filtered Boolean algebras

Projectivity for Boolean algebras is defined as in all other varieties by a diagram condition.

To be read:
A Boolean algebra B is projective iff for all homomorphisms $B \xrightarrow{\varphi} A \xleftarrow{\psi} C$, with ψ surjective, there exists a homomorphism $\varepsilon : B \to C$ such that $\psi \circ \varepsilon = \varphi$.

Putting $A = B$, $\varphi = id$ and letting C be free, we get that each projective Boolean algebra is a retract of a free Boolean algebra. It is easy to prove that this property characterizes projectivity.

It would be desirable to have a characterization of projectivity that refers only to the algebra itself. For other varieties of algebras such intrinsic characterizations have been found. For example, in [14] R. Freese and J.B. Nation characterize projective lattices by four conditions, one of which reads

(∗) *for each $b \in B$ there are two finite sets $U(b) \subseteq \{c \in B : b \leq c\}$*
and $L(b) \subseteq \{c \in B : c \leq b\}$ such that, if $a \leq b$, then $U(a) \cap L(b) \neq \emptyset$.

This condition makes sense for Boolean algebras, too, and for a while it was believed to characterize projective Boolean algebras.[1] It turns out that this is true for Boolean algebras of power at most $\aleph_1$ (cf. 2.2.7) and that all projective Boolean algebras have the property (cf. 2.2.6).

It is remarkable that the class of algebras satisfying (∗) arises in a totally different context. In [55] E.V. Ščepin introduced the class of so-called *openly generated*[2] compact spaces. These are spaces that can be represented as inverse limits of inverse systems $\{X_i, p_j^i, I\}$, where

(1) the partially ordered index set I is σ-complete, i.e. all countable chains $i_1 \leq i_2 \leq \ldots \leq i_n \leq \ldots$ have suprema in I,

(2) the system is continuous, i.e. if $j = \sup J$ exists for some $J \subseteq I$, then X_j is the inverse limit of the restricted system $\{X_j, p_j^i, J\}$,

(3) all X_i are compact and metrizable, and

(4) all bonding maps $p_j^i : X_i \to X_j$ are open.

To make the further explanations precise, we now give two definitions that will be fundamental for the whole work.

A *skeleton* of a Boolean algebra B is a collection $\mathcal{S}$ of subalgebras of B that is closed under unions of chains, i.e. $\bigcup \mathcal{K} \in \mathcal{S}$ whenever $\mathcal{K}$ is a subchain (under $\subseteq$) of $\mathcal{S}$, and absorbing in the sense that for each $X \subseteq B$ there exists some $S \in \mathcal{S}$ such that $X \subseteq S$ and $|S| \leq |X| + \aleph_0$.

A subalgebra A of a Boolean algebra B will be called *relatively complete* (symbolically $A \leq_{rc} B$) if for each $b \in B$ there exists a least element of A above b.

Noticing that relatively complete embeddings are dual to open mappings and that skeletons are (something like) duals of inverse systems, it should be no surprise that the Stone space of a Boolean algebra is openly generated iff the algebra itself has a skeleton consisting of relatively complete subalgebras. As an abreviation we use the expression 'rc-skeleton' and call the algebras that have such skeletons *rc-filtered*. It turns out that (cf. 2.2.3)

A Boolean algebra is rc-filtered iff it has the property (∗).

---

[1] The first author wants to thank M. Ploščica for drawing his attention to this problem.
[2] Also translated as *open generated*.

Comparison of the two classes

Having the property (*), all projective Boolean algebras are rc-filtered. Moreover, the two classes coincide for Boolean algebras of cardinality at most $\aleph_1$.

There are several characterizations of both classes that demonstrate their similarity. In this introduction we just give two such pairs of characterizations. The diagram definition of projectivity can be modified in the following way (due to Širokov, cf. 2.4.3).

The Boolean algebra B is projective iff for all pairs of homomorphisms $B \xrightarrow{\varphi} A \xleftarrow{\psi} C$, with ψ surjective, there exists a mapping $\varepsilon : B \to C$ preserving 0 and $\wedge$ such that $\psi \circ \varepsilon = \varphi$.

The diagram is the same as for projectivity, but ε need not be a homomorphism any more. The counterpart for rc-filtered Boolean algebras reads (cf. 3.2.7):

The Boolean algebra B is rc-filtered iff for all pairs of homomorphisms $B \xrightarrow{\varphi} A \xleftarrow{\psi} C$, with ψ surjective, there exists an order-preserving mapping $\varepsilon : B \to C$ that also preserves disjointness such that $\psi \circ \varepsilon = \varphi$.

On the other hand, the property defining rc-filtered algebras, i.e. the existence of an rc-skeleton, also has a counterpart for projective algebras. It is due to Ščepin and says (cf. 1.3.2(4)):

The Boolean algebra B is projective iff it has a skeleton S such that for each subset $T \subseteq S$ the subalgebra generated by $\bigcup T$ is relatively complete in B.

We only mention one further connection between projective and rc-filtered Boolean algebras (due to A.V. Ivanov, cf. 3.2.6).

The Boolean algebra B is rc-filtered iff λB is projective,

where λB is a Boolean algebra constructed from B in a way explained in section 3.2. Its topological dual is the so-called superextension of the Stone space of B.

Starting from cardinality $\aleph_2$ on, the two classes differ. Much of what follows will be devoted to the construction of rc-filtered Boolean algebras that are not projective. Most of them have additional properties which show that they are non-projective 'in a strong sense'. Let us list the most interesting of these. *There are rc-filtered Boolean algebras which are*

(1) *not embeddable into a free Boolean algebra* (cf. 3.3.11),

(2) *not projective but relatively complete subalgebras of free Boolean algebras* (cf. 3.4.7),

(3) *not co-complete to a projective Boolean algebra* (cf. 6.3.2).

Moreover, Fuchino proved (unpublished, cf. 6.4.2) that

there are $2^{\aleph_2}$ pairwise non-isomorphic rc-filtered Boolean algebras of power $\aleph_2$.

Let us mention that among them there are only $2^{\aleph_1}$ projective Boolean algebras (by a result of Koppelberg's [35] not reproduced here).

The class of rc-filtered Boolean algebras

In chapter 2 the class of rc-filtered Boolean algebras is studied in a rather systematic way. Let us just mention two results about the behaviour of rc-filtered Boolean algebras with respect to various operations (cf. 2.2.8 and 2.3.1).

Relatively complete subalgebras of rc-filtered Boolean algebras remain rc-filtered.

If B can be written as the union of a well-ordered continuous chain $(B_\alpha)_{\alpha < \lambda}$ of rc-filtered subalgebras such that $B_\alpha \leq_{rc} B_\beta$ for all $\alpha < \beta$, then B is itself rc-filtered.

We also study cardinal functions on rc-filtered Boolean algebras and their subalgebras. It turns out that with respect to the most popular functions these algebras behave like free ones. More precisely (cf. 2.7.10), *if B is a subalgebra of an rc-filtered Boolean algebra, then*

$$\pi\chi = \mathrm{ind} = \pi = \mathrm{Irr} = t = s = \chi = \mathrm{hL} = \mathrm{hd} = \mathrm{Inc} = \mathrm{h\text{-}cof} = |B|$$
$$\mathrm{V\!I}$$
$$d$$
$$\mathrm{V\!I}$$
$$\mathrm{Depth} = \mathrm{Length} = c = \aleph_0.$$

Co-completeness and weak projectivity

Two Boolean algebras will be called co-complete if they have isomorphic completions. Chapter 5 is devoted to the class of Boolean algebras that are co-complete with projective Boolean algebras. In want of a better name, we call them *weakly projective*. The main results are characterizations of weak projectivity. Two highlights from that chapter are theorem 5.3.11 saying that

every subalgebra of a projective Boolean algebra is weakly projective

and theorem 5.2.2, which determines weakly projective Boolean algebras up to co-completeness as the at most countable products of free Boolean algebras.

In the context of co-completeness the relevant type of embedding is called *regular*. We write $A \leq_{reg} B$ iff $sup^A M = sup^B M$ for each $M \subseteq A$ such that $sup^A M$ exists (i.e. the identical mapping preserves all infinite suprema existing in A).

It is rather easy to prove that each Boolean algebra that is co-complete to an rc-filtered one is 'regularly filtered', i.e. has a skeleton consisting of regular subalgebras. Whether the converse is also true, remains an open problem.

Adequate pairs

At the beginning of the investigations in connection with his 'spectral theorem' Ščepin defined the notion of a class $\mathcal{X}$ of compact spaces being adequate for a class Φ of continuous mappings. The following is a slightly modified Boolean algebraic version of this concept.

Let $\mathcal{B}$ be a class of Boolean algebras and $\mathcal{E}$ a class of embeddings. We write $A \leq_{\mathcal{E}} B$ to express that $A \leq B$ belongs to $\mathcal{E}$. We call the classes $\mathcal{B}$ and $\mathcal{E}$ *adequate* if the following conditions are satisfied.

(A1) Every algebra in $\mathcal{B}$ has a skeleton $\mathcal{S}$ such that $S \leq_{\mathcal{E}} T$ for all $S \leq T$ belonging to $\mathcal{S}$.

(A2) If $(B_\alpha)_{\alpha < \lambda}$ is a well-ordered continuous chain of Boolean algebras belonging to $\mathcal{B}$ such that $B_\alpha \leq_{\mathcal{E}} B_\beta$ for all $\alpha < \beta$, then $B = \bigcup_{\alpha < \lambda} B_\alpha$ belongs to $\mathcal{B}$ and $B_\alpha \leq_{\mathcal{E}} B$, for all $\alpha < \lambda$.

The class of rc-filtered Boolean algebras is adequate for the class of relatively complete embeddings and the class of regularly filtered Boolean algebras turns out to be adequate for the class of regular embeddings. In both cases condition (A1) is taken as definition and (A2) proved from it. The same is true for the third pair that we study in chapter 4. It is defined in terms of σ-embeddings, where $A \leq_\sigma B$ if, for each $b \in B$, the ideal $\{a \in A : a \leq b\}$ is countably generated. The results parallel those for the other two pairs.

Let us mention here that there are also classes of embeddings which are adequate for the class of projective and weakly projective Boolean algebras. Appropriately, they are called projective and weakly projective embeddings. The definitions are more complicated than in the above cases and can be found in sections 1.5 and 5.3, respectively.

Functors

In chapter 3 we consider three constructions that are, in fact, covariant functors of the category of Boolean algebras into itself: λ, exp, and SP^2. Their main purpose is to prove Ivanov's theorem and to construct the examples (1) and (2) mentioned on page 5 above. The algebras in question are $exp \, \mathrm{Fr} \, \omega_2$ and $SP^2(\mathrm{Fr} \, \omega_2)$, respectively, where $\mathrm{Fr} \, X$ denotes the free Boolean algebra on the set X of generators.

The examples show that the class of projective algebras is not closed under the functors exp and SP^2. This observation naturally leads to more general question of closedness under these functors. We concentrate on exp, where the principal results are 3.3.10, 3.3.6, and 5.4.5:

> $exp\,A$ is projective iff A is projective and $|A| \leq \aleph_1$.
>
> $exp\,A$ is rc-filtered iff A is rc-filtered.
>
> $exp\,A$ is weakly projective iff A is weakly projective.

It should be mentioned here that in the topological setting there is a theory of so-called normal functors and that (slight modifications of) the above and other results below hold for these in general. Our restriction[3] to exponentials has several reasons. First of all, exp is *the* typical normal functor. In that sense we dont loose much. Moreover, the proofs are technically more transparent for exponentials than in the general case. Finally, the very definition of a normal functor becomes rather clumsy and unnatural if translated into the Boolean algebraic language. The reader who wants to know more is referred to [55].

Set-theoretic appendix

The results in the main text are all obtained in ZFC by orthodox topological and algebraic methods. The appendix written by Sakaé Fuchino demonstrates another method to obtain results in ZFC. It uses elementary submodels of models of set theory and was first applied to topological questions independently by I. Bandlow and A. Dow.

Moreover, the appendix contains a number of recent independence results mostly due to Fuchino himself concerning rc-filtered Boolean algebras which answer some questions of Ščepin from [55].

Prerequisites and notation

We present all definitions and results in the language of Boolean algebras and the reader is supposed to have some experience with them. Our standard reference will be the *Handbook of Boolean algebras,* in particular its first volume [34]. Whenever possible we quote results from there. This is rather unjust to the original authors, but, hopefully, convenient for the reader.

Modulo the Handbook the text is more or less self-contained. Some 'Digressions' contain results that shed additional light on what is in the main text. Some of them are quoted without proof and qualified as 'Informations'. None of these results will be used in later proofs.

With very few exceptions we use standard notation, i.e. that of the Handbook. Let us dwell on some points that may differ from what the reader is used to.

Boolean operations

As a rule we consider Boolean algebras as complemented distributive lattices, i.e. with the fundamental operations of intersection = meet, union = join, and

---

[3]But notice that SP^2 is also normal.

complementation. We stick to the good old symbols $\wedge$, $\vee$, and $-$ (the latter is officially unary; but $a - b$ stands for $a \wedge -b$).

If $F = \{a_1, \ldots a_n\}$ is a finite set of elements of the Boolean algebra A, we alternatively write $\bigvee F$, $a_1 \vee \ldots \vee a_n$ or $\bigvee_{i=1}^n a_i$ to denote its join. The a_i are then called 'joinands'. If the set F is infinite, we still write $\bigvee F$ for its supremum (if it exists). The elements of F will still be 'joinands'. If several algebras are considered at the same time, it makes sense to indicate in which algebra the supremum is taken and we write $\bigvee^A F$. Similarly for finite and infinite meets and 'meetands'.

Sometimes (and then we emphasize this) it will be convenient to consider Boolean algebras as linear algebras (i.e. vector spaces with a multiplication) over the field $\mathbf{F}_2$ with two elements. That is why we use $\wedge, \vee$ and $-$ to denote the lattice-theoretic operations and $+$ and $\cdot$ for the ring-theoretic ones. The connection is well known:

$$a \cdot b = a \wedge b, \quad a + b = (a \vee b) - (a \wedge b), \quad a \vee b = a + b + a \cdot b, \quad -a = 1 + a.$$

Subalgebras and embeddings

$A \leq B$ means that A is a subalgebra of B. Formally, an *embedding* (sometimes also called *extension*) is a pair (A, B) such that $A \leq B$. We usually suppress the parentheses and write, e.g., 'let $A \leq B$ be an embedding...'. The more general concept of embedding hardly ever occurs in what follows and if it does, it will be called an injective homomorphism.

For $A \leq B$ and $b \in B$ we let $A \upharpoonright b$ denote the ideal $\{a \in A : a \leq b\}$ of A. If $b \in A$ then $A \upharpoonright b$ is a principal ideal, which can also be considered as a Boolean algebra, the so-called factor algebra of A corresponding to b. It will be clear from the context if we mean the factor algebra.

For a subset $X \subseteq A$ of a Boolean algebra, $\langle X \rangle_A$ denotes the subalgebra of A generated by X. Usually A will be clear from the context and we write $\langle X \rangle$ only. If C is a subalgebra of A and $X \subseteq A$, we sometimes write $C(X)$ instead of $\langle C \cup X \rangle$. If $X = \{x_1, \ldots, x_n\}$ is finite, this notation becomes $C(x_1, \ldots, x_n)$.

We shall often meet subalgebras of the form $\langle B \cup C \rangle_A$, where B and C are subalgebras of A. The elements of $\langle B \cup C \rangle$ have a particularly simple description, namely $\bigvee_{i=1}^n b_i \wedge c_i$, where $b_i \in B$ and $c_i \in C$. This trivial fact often makes life easier and will be tacitly used throughout.

Free products

By $A \otimes B$ we denote the free product of A and B. (cf. subsection 11.1 of [34], where the notation $A \oplus B$ is used). We find it more illuminating to denote its canonical generators by $a \otimes b$ (instead of the $e_A(a) \wedge e_B(b)$ of the *Handbook*). So, each element of $A \otimes B$ can be written in the form $\bigvee_{i=1}^n a_i \otimes b_i$ for some $a_1, \ldots a_n \in A$ and $b_1, \ldots b_n \in B$. The characteristic property of free products is expressed by the following fact.

For each pair $\varphi : A \to C$ and $\psi : B \to C$ of homomorphisms there is a unique homomorphism $\varphi \otimes \psi : A \otimes B \longrightarrow C$ such that $(\varphi \otimes \psi)(a \otimes b) = \varphi(a) \wedge \psi(b)$.

The free product of an infinite family will be written as $\bigotimes_{i \in I} A_i$.

Sikorski's Extension Criterion

The following theorem (5.5 in [34]) will be used in several places and it seems appropriate to formulate it once in all detail.

Theorem 0.0.1 *Assume X generates the Boolean algebra A and φ maps X into a Boolean algebra B. For φ to extend to a homomorphism $A \to B$ it is necessary and sufficient that for all $x_1, \ldots x_n \in X$ and all $\varepsilon_1, \ldots, \varepsilon_n \in \{+1, -1\}$ if $\varepsilon_1 x_1 \wedge \ldots \wedge \varepsilon_n x_n = 0$ in A, then $\varepsilon_1 \varphi(x_1) \wedge \ldots \wedge \varepsilon_n \varphi(x_n) = 0$ in B.*

Here $+1x$ means x and $-1x$ is $-x$. In practice the given condition often splits into three (collecting 'positive and negative' elements on different sides).

$$x_1 \wedge \ldots \wedge x_n = 0 \implies \varphi(x_1) \wedge \ldots \wedge \varphi(x_n) = 0,$$
$$x_1 \vee \ldots \vee x_n = 1 \implies \varphi(x_1) \vee \ldots \vee \varphi(x_n) = 1,$$
and
$$x_1 \wedge \ldots \wedge x_m \leq x_{m+1} \vee \ldots \vee x_n \implies$$
$$\varphi(x_1) \wedge \ldots \wedge \varphi(x_m) \leq \varphi(x_{m+1}) \vee \ldots \vee \varphi(x_n).$$

Set theory

Our notation is standard. As usual, we consider cardinal numbers as special ordinals. In particular, $\aleph_\alpha$ and ω_α denote the same object, considered under different aspects. It will be convenient to use the notation $|X|$ to denote the maximum of $\aleph_0$ and the cardinality of X, i.e. $|X|$ is always infinite.

Very little set theory is needed in the main text. In sections 2.10 and 6.4 we use stationary sets and some of their basic properties. Everything we need to know about these sets can be found in all modern standard texts. It is also contained in J. D. Monk's *Appendix on set theory* to volume 3 of the *Handbook* [34] (which the reader is likely to use anyway). The same is true of the (easiest version) of the Δ-Lemma, which occurs several times in the text. As with Sikorski's Theorem above, we feel obliged to once formulate it in full detail:

Theorem 0.0.2 *If κ is a regular uncountable cardinal and $(X_\alpha)_{\alpha < \kappa}$ is a family of finite sets, then there exists a subset $K \subseteq \kappa$ and a finite set Y such that $|K| = \kappa$ and $(X_\alpha)_{\alpha \in K}$ is a Δ-system with kernel Y, i.e. $X_\alpha \cap X_\beta = Y$ for all distinct $\alpha, \beta \in K$.*

On one occasion we also need the analogous statement for families of more than $2^{\aleph_0}$ countable sets. An unorthodox proof of the general form is contained in Fuchino's appendix (cf. A.1.13).

Chapter 1

Setting the stage

The first two sections of this chapter will be devoted to some necessary technical preparations. After that we shortly review projectivity in the class of Boolean algebras. Part of these results will be used later in the text. The main purpose of the review is, however, to set the stage for the generalizations to come.

In what follows A, B, C always denote Boolean algebras if not explicitly stated otherwise. Needless to say that we follow common practice and denote an algebra and its underlying set by the same character.

1.1 Clubs, skeletons and filtrations

In this section we introduce the three concepts mentioned in its title and prove some very elementary properties.

Definition 1.1.1 A *club* on a set X is a collection of *countable* subsets of X which is *closed* under unions of countable chains[1] (with respect to $\subseteq$) and *unbounded* in the sense that each countable subset of X is contained in some member of the club. $\square$

It is probably a good first exercise to verify that the collection of all countable subalgebras of a Boolean algebra A is a club on A. Skeletons are similar to clubs of subalgebras but with the countability conditions modified.

Definition 1.1.2 Let A be Boolean algebra and $\mathcal{S}$ a collection of subalgebras of A.

(1) We say that $\mathcal{S}$ is *closed* if the union of every subchain of $\mathcal{S}$ belongs to $\mathcal{S}$.

(2) We say that $\mathcal{S}$ is *absorbing* if for each $C \leq A$ there is some $S \in \mathcal{S}$ such that $C \leq S$ and[2] $|S| \leq |C|$.

---

[1] i.e. linearly ordered subsets
[2] Recall that we agreed to make $|X|$ always infinite.

(3) S is called a *skeleton* of A if it is both, closed and absorbing.

(4) A skeleton S is called *additive* if $\langle \bigcup \mathcal{K} \rangle_A \in S$ for all subsets $\mathcal{K} \subseteq S$. $\square$

As every chain contains a well-ordered cofinal subchain, it is sufficient to demand that a skeleton be closed under unions of well-ordered chains. Notice that the countable members of each skeleton form a club.

Lemma 1.1.3 *If S and T are skeletons of A, then $S \cap T$ and $S \cup T$ are also skeletons of A. If all S_n are skeletons of A, then $\bigcap_{n<\omega} S_n$ is a skeleton of A. The same is true for clubs.*

Proof. $S \cap T$ is obviously closed. To see that it is also absorbing, consider any $C \leq A$. Using that S and T are skeletons it is easy to find by induction

$$C \leq S_0 \leq T_0 \leq S_1 \leq T_1 \leq S_2 \leq \ldots$$

all of power $|C|$ such that $S_i \in S, T_i \in T$. Then $\bigcup_{i<\omega} S_i = \bigcup_{i<\omega} T_i$ belongs to $S \cap T$, has power $|C|$ and absorbs C.

The assertion about the union is obvious. Notice that it is sufficient that one of S or T is a skeleton and the other is only closed.

Now let a sequence $(S_n)_{n<\omega}$ of skeletons be given. It is obvious that $\bigcap_{n<\omega} S_n$ remains closed. To prove absorption consider any $C \leq A$. Inductively we pick $C \leq S_0 \leq S_1 \leq \ldots \leq S_n$ such that $|S_n| \leq |C|$ and $S_n \in S_0 \cap S_1 \cap \ldots \cap S_n$. This is possible because, by the already proved, finite intersections of skeletons remain absorbing. Then $C \leq S = \bigcup_{n<\omega} S_n$ which has power $|C|$ and belongs to $\bigcap_{n<\omega} S_n$. Indeed, for all n, $S = \bigcup_{m=n}^{\infty} S_m \in S_n$, by closedness of S_n.

The assertions about skeletons are proved. The same argument works for clubs as well. $\square$

The following lemma will be useful later on. We prove it here because it is a good exercise to get familiar with the definition.

Lemma 1.1.4 *Assume $A \leq B$ and let S be a skeleton of A. Then the collection $T = \{T \leq B : T \cap A \in S\}$ is a skeleton of B.*

Proof. It should be clear that T is closed. To see that it is also absorbing, consider any $C \leq B$. Choose $S_0 \in S$ such that $C \cap A \leq S_0$ and $|S_0| \leq |C|$. Then choose by induction $S_n \leq S_{n+1} \in S$ such that $|S_{n+1}| \leq |C|$ and $\langle C \cup S_n \rangle \cap A \leq S_{n+1}$. Putting $S = \bigcup_{n<\omega} S_n \in S$ we get

$$S \leq \langle C \cup S \rangle \cap A = \bigcup_{n<\omega} \langle C \cup S_n \rangle \cap A \leq \bigcup_{n<\omega} S_{n+1} = S$$

So, $\langle C \cup S \rangle \in T$. Moreover, $C \leq \langle C \cup S \rangle$ and $|\langle C \cup S \rangle| \leq |C|$. $\square$

In the other direction things don't go that smoothly. Assume $A \leq B$ and let S be a skeleton of B. If we want to manufacture a skeleton of A, the first idea is to try $\{A \cap S : S \in \mathcal{S}\}$. This family is easily seen to be absorbing, but it need not be closed. The reason is that there may be 'ill-indexed' chains whose union does not belong to the family ($A \cap S \leq A \cap T$ may hold, whereas $S \leq T$ does not). There are two ways to overcome this problem. One can prove that the family in question contains a skeleton. The other way is to observe that $\{A \cap S : S \in \mathcal{S}\}$ retains enough continuity for most applications. This becomes apparent if we write A_S for $A \cap S$ and consider the indexed family $(A_S)_{S \in \mathcal{S}}$ with the index set $\mathcal{S}$ partially ordered by inclusion. Then $A_K = \bigcup_{S \in \mathcal{K}} A_S$ for the least upper bound $K = \bigcup \mathcal{K}$ of every subchain $\mathcal{K}$ of $\mathcal{S}$. Notice also that $(A_S)_{S \in \mathcal{S}}$ exhausts all of A and that the partially ordered set $\mathcal{S}$ is directed, i.e. for all $S, S' \in \mathcal{S}$ there is some $T \in \mathcal{S}$ such that $S \subseteq T$ and $S' \subseteq T$. Both properties follow from $\mathcal{S}$ being absorbing.

The essential properties of the family $(A_S)_{S \in \mathcal{S}}$ are reflected in the following

Definition 1.1.5 Let $(I, \leq)$ be a partially ordered set. An indexed family $(A_i)_{i \in I}$ of subalgebras of a Boolean algebra A will be called a *filtration* of A if the following conditions are satisfied.

(1) $(I, \leq)$ is (upwards) directed, i.e. for all $i, j \in I$ there exists some $k \in I$ such that $i \leq k$ and $j \leq k$.

(2) If $i \leq j$, then $A_i \leq A_j$.

(3) $A = \bigcup_{i \in I} A_i$

(4) If $K \subseteq I$ is a chain and $i = \sup K$ exists, then $A_i = \bigcup_{k \in K} A_k$. □

Condition (4) will sometimes be referred to as the continuity of the filtration. Notice that we do not require that $\sup K$ exists. Demanding that certain chains in I have suprema, is an extra condition on the filtration. Usually we demand that all countable chains have suprema in I and call the filtration *σ-complete* then.

An important special case deserves special mentioning and some special notation. It occurs if the index set is well-ordered. We then usually write $(A_\alpha)_{\alpha < \gamma}$, where γ is an ordinal (most often even a cardinal) and speak of a *well-ordered filtration*. If we want to stress that a filtration is not (necessarily) well-ordered, we call it *ramified*.

It is clear that clubs of subalgebras and skeletons (considered as indexed by themselves) are σ-complete filtrations. A well-ordered filtration $(B_\alpha)_{\alpha < \lambda}$ is σ-complete iff $cf(\lambda) \neq \omega$.

The following two observations are almost trivial, but very useful.

Observation 1.1.6 *Let $(B_i)_{i \in I}$ be a σ-complete filtration of some Boolean algebra B and consider some subset $M \subseteq B$.*

(1) *If M is countable, then $M \subseteq B_i$ for some $i \in I$.*

(2) *If* $|M| \leq \aleph_1$, *then there exists a continuous[3] chain* $(i_\alpha)_{\alpha < \omega_1}$ *of indices such that* $M \subseteq \bigcup_{\alpha < \omega_1} B_{i_\alpha}$.

Proof. Assertion (1) being even simpler, we confine ourselves to (2). We may enumerate M by the ordinals $< \omega_1$: $M = \{m_\alpha : \alpha < \omega_1\}$. As the B_j cover B, for each α there is $j_\alpha \in I$ such that $m_\alpha \in B_{j_\alpha}$. Now we define i_α by induction starting with an arbitrary i_0. At limit stages we use σ-completeness to put $i_\beta = \sup_{\alpha < \beta} i_\alpha$. At successor stages we use that I is directed, which makes it possible to choose $i_{\alpha+1} \geq i_\alpha, j_\alpha$. $\Box$

Let us return to skeletons. Here is how they usually arise.

Definition 1.1.7 Let f be a mapping from a Boolean algebra A to the (at most) countable subsets of A, in symbols $f : A \to [A]^{\leq \aleph_0}$. The set

$$S_f = \{S \leq A : s \in S \text{ implies } f(s) \subseteq S\}$$

will be called the *fixed-point skeleton* of f. $\Box$

The reader will have no difficulty to verify that S_f is, indeed, a skeleton. With the obvious modification in mind we also speak of the fixed-point skeleton of one or several mappings $A \to A$.

Topological duality: Our filtrations correspond to continuous spectra (i.e. inverse systems of topological spaces), well-oredered filtrations correspond to 'transfinite spectra' and the topological analogs of our skeletons are called 'lattices' in Ščepin's papers.

1.2 Three types of embeddings

One way of classifying embeddings is to look at the properties of ideals that are preserved. If $A \leq B$ is an embedding and $I \subseteq B$ is an ideal of B, then $A \cap I$ is an ideal of A which may or may not share a property with I. A local version of this principle, which often yields the same class of embeddings, is obtained if only principal ideals of B are considered: $A \leq B$ falls in the class if the ideal $A \upharpoonright b = \{a \in A : a \leq b\}$ has the corresponding property for all $b \in B$. Below we consider three properties of ideals: being principal, being regular, and being countably generated. They yield the classes of relatively complete, regular, and σ-embeddings, respectively.

Definition 1.2.1 Assume $A \leq B$. We say that A is *relatively complete* in B and write $A \leq_{rc} B$ if $A \upharpoonright b$ is principal for all $b \in B$. $\Box$

The prase 'relatively complete' is motivated by the equivalent condition that $\bigvee^A \{a \in A : a \leq b\}$ exists for each $b \in B$. Switching to complements it is clear that $A \leq_{rc} B$ implies the existence of $\bigwedge^A \{a \in A : b \leq a\}$ for each $b \in B$.

---

[3]i.e. $i_\beta = \sup_{\alpha < \beta} i_\alpha$, for all limit ordinals β.

The mapping sending b to that infimum, i.e. to the least element of A above b, will be called the *projection* associated with $A \leq_{rc} B$ and denoted by q_A^B or often only by q. This differs a little from standard terminology (cf. [35]), where $p_A^B : B \to A$ with $p_A^B(b) = \max\{a \in A : a \leq b\}$ is called the projection. The reason we prefer q is that it occurs more naturally than p in most arguments involving relative completeness. To distinguish the two we sometimes call q the *upper* and p the *lower* projection. Besides, the two mappings are dual to each other: $q(b) = -p(-b)$.

We now list some properties of relatively complete embeddings that are needed in the text below. Althogh the results are well-known, proofs will be given or sketched, for completeness sake. Further information can be found in [34] and [35], where most of the following material is taken from. We start with some immediate consequences of the definition.

Observations 1.2.2

(1) *Complete, in particular finite, subalgebras are relatively complete.*

(2) *The projections q_A^B are no Boolean algebra homomorphisms, but they are monotone, i.e. $b_1 \leq b_2$ implies $q_A^B(b_1) \leq q_A^B(b_2)$. The same is true for the lower projections.*

(3) *If $A \leq_{rc} B \leq D$ and $A \leq_{rc} C \leq D$ then q_A^B and q_A^C agree on $B \cap C$.*

The following assertion makes it easier to test whether a given embedding $A \leq B$ is relatively complete.

Lemma 1.2.3 *Assume $A \leq B$ and let $D \subseteq B$ be such that each element of B is equal to a finite union of elements of D. If there is a minimal element of A above each $d \in D$, then $A \leq_{rc} B$.*

Proof. Consider an arbitrary $b \in B$ and write it as $b = d_1 \vee \ldots \vee d_n$ with $d_i \in D$. Let $d_i \leq a_i \in A$ be minimal. Then $b \leq a = a_1 \vee \ldots \vee a_n$. To see that a is minimal above b, consider any $b \leq c \in A$. Then $d_i \leq c$, hence $a_i \leq c$, for all i. It follows that $a \leq c$, as desired. $\square$

Here is another equally trivial, but useful test.

Lemma 1.2.4 *Assume $A \leq B$ and let $E \subseteq A$ be such that each element of A is equal to a finite union of elements of E. Consider $0 < b \leq a$ with $a \in A$ and $b \in B$. Then a is minimal in A above b iff each $e \in E$ that intersects a also intersects b, i.e. iff $b \wedge e = 0$ implies $a \wedge e = 0$, for all $e \in E$.*

Proof. Let a be minimal and $a \wedge e \neq 0$. Assuming $b \wedge e = 0$, we would get $b \leq a - e < a$, contradicting minimality.

The other way round we consider $b \leq a' \in A$. Assuming $a - a' \neq 0$, we get some $0 < e \in E$ such that $e \leq a - a'$. This e intersects a but not a' let alone b. $\square$

As a first simple but important application of the two lemmas above we make the following

Observation 1.2.5 *If* B *is generated by* $A \cup C$, *where* A *and* C *are independent,*[4] *then* $A \leq_{rc} B$. *In particular, if* F *is a free Boolean algebra freely generated by the set* X *and* $Y \subseteq X$, *then* $\langle Y \rangle \leq_{rc} F$.

Proof.   Each non-zero $b \in B$ can be written as a union of elements of

$$D = \{a \wedge c : 0 < a \in A, \ 0 < c \in C\}.$$

By 1.2.3, it is sufficient to establish that a is minimal in A above $0 < d = a \wedge c$. By 1.2.4 (with $E = A$), this follows from

$$a' \wedge (a \wedge c) = 0 \quad \Rightarrow \quad a' \wedge a = 0$$

for all $a' \in A$. But this implication is an immediate consequence of independence. The second assertion is the special case $B = F$, $A = \langle Y \rangle$, $C = \langle X \setminus Y \rangle$. $\square$

As an exercise the reader may now verify

Observation 1.2.6 *If* $A_i \leq_{rc} B_i$, *then* $A_1 \otimes A_2 \leq_{rc} B_1 \otimes B_2$.

The sets that make life easier here are, of course,

$$D = \{b_1 \otimes b_2 : b_1 \in B_1; \ b_2 \in B_2\}$$
$$\text{and}$$
$$E = \{a_1 \otimes a_2 : a_1 \in A_1; \ a_2 \in A_2\}.$$

One important property of classes of embeddings is transitivity. Relatively complete embeddings enjoy it in a strong way.

Lemma 1.2.7 *Assume* $A \leq B \leq C$.

(1) *If* $A \leq_{rc} B$ *and* $B \leq_{rc} C$, *then* $A \leq_{rc} C$.

(2) *If* $A \leq_{rc} C$, *then* $A \leq_{rc} B$.

(3) *If* $A \leq_{rc} C$ *and* B *is finitely generated over* A, *then* $B \leq_{rc} C$.

Proof.   (1) A straightforward verification shows that $q_A^B \circ q_B^C$ acts as the projection $C \to A$. (2) is obvious, $q_A^C \restriction B$ is the desired projection.

(3) is a little harder. A simple induction reduces the assertion to the case $B = A(x) = \langle A \cup \{x\} \rangle$. Then each $b \in B$ can be written as $b = (a_1 \wedge x) \vee (a_2 - x)$.

Let $c \in C$ be given. Let a_1, a_2 be the minimal elements of A above $c \wedge x$ and $c - x$, respectively. Then $B \ni b = (a_1 \wedge x) \vee (a_2 - x) \geq c$ and this b is even minimal above c in B. Indeed, $c \leq b' = (a_1' \wedge x) \vee (a_2' - x)$ implies

$$c \wedge x \leq a_1' \wedge x \leq a_1', \text{ from which } a_1 \leq a_1'$$

and

$$c - x \leq a_2' - x \leq a_2', \text{ from which } a_2 \leq a_2'.$$

---

[4] i.e. $a \wedge c \neq 0$ for all non-zero $a \in A$ and $c \in C$.

So, $b \leq b'$, as was to be shown. $\square$

Notice that the finite generation in (3) is indespensable. For, otherwise, from $2 \leq B \leq C$ and $2 \leq_{rc} C$ we would get $B \leq_{rc} C$ for arbitrary embeddings. Still on the subject of transitivity we have

Lemma 1.2.8 *If $\mathcal{A}$ is a chain of Boolean algebras such that either $A \leq_{rc} A'$ or $A' \leq_{rc} A$ for all $A, A' \in \mathcal{A}$, then $A \leq_{rc} \bigcup \mathcal{A}$, for all $A \in \mathcal{A}$.*

Proof. The mapping

$$\bigcup \{q_A^{A'} : A \leq A' \in \mathcal{A}\}$$

is easily seen to be the desired projection. $\square$

From the lemmas we get transitivity for arbitrary filtrations.

Corollary 1.2.9 *Let $(A_\alpha)_{\alpha < \gamma}$ be a well-ordered filtration of A.*

(1) *If $A_\alpha \leq_{rc} A_{\alpha+1}$ for all α, then $A_\alpha \leq_{rc} A_\beta \leq_{rc} A$ for all $\alpha \leq \beta < \gamma$.*

(2) *For all $\alpha < \beta < \delta$ and the projections $A_\delta \xrightarrow{q_\beta^\delta} A_\beta \xrightarrow{q_\alpha^\beta} A_\alpha$ and $A_\delta \xrightarrow{q_\alpha^\delta} A_\alpha$ it holds that*
$$q_\alpha^\delta = q_\alpha^\beta \circ q_\beta^\delta \quad \text{and} \quad p_\alpha^\delta \mid A_\beta = p_\alpha^\beta.$$

In section 1.4 below, we need the following technical

Lemma 1.2.10 (cf. 3.2 in [35])
If $A \leq_{rc} B$ then $I = \{q_A^B(b) \wedge q_A^B(-b) : b \in B\} \subseteq A$ is an ideal.

Proof. We denote the projection q_A^B simply by q. In the following claims a and b stand for elements of A and B, respectively.

CLAIM 1. $q(a \wedge b) = a \wedge q(b)$

Indeed, from $b \leq q(b)$ we get $a \wedge b \leq a \wedge q(b) \in A$, hence $q(a \wedge b) \leq a \wedge q(b)$. For the opposite inequality notice that $a \wedge b \leq q(a \wedge b)$ implies $b \leq q(a \wedge b) \vee -a \in A$. Hence $q(b) \leq q(a \wedge b) \vee -a$ and $a \wedge q(b) \leq q(a \wedge b)$.

CLAIM 2. $q(a \vee b) = a \vee q(b)$

Indeed, from $a \vee b \leq a \vee q(b) \in A$ we get $q(a \vee b) \leq a \vee q(b)$. The opposite inequality follows from the obvious (q is monotone) $a \leq q(a \vee b)$ and $q(b) \leq q(a \vee b)$.

As an abbreviation we put $i(b) = q(b) \wedge q(-b)$. Then claims 1 and 2 yield

$$\begin{aligned}
i(a \wedge b) &= q(a \wedge b) \wedge q(-(a \wedge b)) \\
&= a \wedge q(b) \wedge q(-a \vee (a - b)) \\
&= a \wedge q(b) \wedge [-a \vee (a \wedge q(-b))] \\
&= a \wedge q(b) \wedge q(-b) \\
&= a \wedge i(b).
\end{aligned}$$

It follows immediately that I is downwards closed: $a \leq i(b)$ implies $i(a \wedge b) = a \wedge i(b) = a$.

We finish by proving that I is closed under disjoint unions. Consider $a = i(b)$ and $a' = i(b')$ such that $a \wedge a' = 0$.

CLAIM 3. $a \vee a' = i(c)$ *for* $c = (a \wedge b) \vee (a' \wedge b')$.

From $c \leq a \vee a'$, we get $i(c) \leq q(c) \leq a \vee a'$. By symmetry, it will be sufficient to verify $a \leq i(c)$, i.e. $a \leq q(c)$ and $a \leq q(-c)$. Now $a = i(b) \leq q(b)$ implies $a = q(a \wedge b) \leq q(c)$. The disjointness of a and a' yields $(a - b) \wedge c = 0$, hence $a - b \leq -c$. Therefore, $a = i(b) \leq q(-b)$ implies $a = q(a - b) \leq q(-c)$. $\square$

Now we pass to regular embeddings.

Definition 1.2.11 Let M be a subset of a Boolean algebra A. The *disjoint complement* of M (with respect to A) is the set

$$M^d = \{a \in A : a \wedge m = 0 \text{ for all } m \in M\}. \quad \square$$

Next we list some easy observations that will be (often tacitly) used when dealing with disjoint complements.

Observations 1.2.12 *Let M and N be subsets of some Boolean algebra A.*

(1) $M \cap M^d \subseteq \{0\}$

(2) M^d *is an ideal of A*

(3) $M \subseteq N$ *implies* $N^d \subseteq M^d$, *hence* $M^{dd} \subseteq N^{dd}$.

(4) $M \subseteq M^{dd}$

(5) $M^d = M^{ddd}$

Assertions (1) – (4) should be obvious. To prove (5), one applies (3) to (4), yielding $M^{ddd} \subseteq M^d$. The reverse inclusion is a special case of (4). $\square$

The following assertion will be needed in section 5.3 only.

Lemma 1.2.13 *If I and J are ideals of the Boolean algebra A, then*

$$(I \cap J)^{dd} = I^{dd} \cap J^{dd}.$$

Proof. We first show that

(∗) $I \cap J^{dd} \subseteq (I \cap J)^{dd}$

Let $i \in I \cap J^{dd}$ and $b \in (I \cap J)^d$ be given. We have to show $i \wedge b = 0$. Take $j \in J$. Then $(b \wedge i) \wedge j = b \wedge (i \wedge j) = 0$, because $i \wedge j \in I \cap J$ (here we need that the two sets are ideals). So, $b \wedge i \in J^d$. But also $i \in J^{dd}$, hence $b \wedge i \in J^d \cap J^{dd} = \{0\}$, as desired.

Applying (*) with I^{dd} in place of I yields

(+) $I^{dd} \cap J^{dd} \subseteq (I^{dd} \cap J)^{dd}$.

Interchanging I and J in (*) we get $I^{dd} \cap J \subseteq (I \cap J)^{dd}$ and, applying observations (3) and (5) above,

(++) $(I^{dd} \cap J)^{dd} \subseteq (I \cap J)^{dddd} = (I \cap J)^{dd}$.

Combining (+) and (++), we obtain $I^{dd} \cap J^{dd} \subseteq (I \cap J)^{dd}$. The reverse inclusion is a simple consequence of the monotonicity of ^{dd}. $\square$

Definition 1.2.14 An ideal I is called *regular* if $I = I^{dd}$. We say that A is a *regular* subalgebra of B and write $A \leq_{reg} B$ if $I \cap A$ is regular, for all regular ideals $I \subseteq B$. $\square$

Observations 1.2.15 *Let I be an ideal of the Boolean algebra A.*

(1) I is regular iff $I^{dd} \subseteq I$.

(2) If I is regular, then $I = M^d$ for some $M \subseteq A$ consisting of pairwise disjoint elements.

(3) If A satisfies the ccc[5] and I is regular, then $I = M^d$ for some countable $M \subseteq A$.

Indeed, (1) is obvious and (3) follows immediately from (2). To prove the latter, let I be regular. Use Zorn's Lemma to choose a maximal subset $M \subseteq I^d$ consisting of pairwise disjoint elements. Then $I = I^{dd} \subseteq M^d$ is clear. If $a \notin I = I^{dd}$, then $a \wedge b > 0$ for some $b \in I^d$, by maximality of M, we must have $a \wedge b \wedge m \neq 0$ for some $m \in M$. So, $a \notin M^d$, as desired. $\square$

It follows from 1.2.12(2,5) that M^d is always a regular ideal. As an exercise the reader may prove that M^{dd} is the smallest regular ideal containing M. If I is an ideal, then I^{dd} is sometimes called the *regularization* of I.

There are several characterizations of regular embeddings that will be useful below. They are listed in the following

Proposition 1.2.16 *For an embedding $A \leq B$ the following assertions are equivalent.*

(1) $A \leq_{reg} B$

(2) $A \restriction b$ is regular for each $b \in B$.

(3) For each subset M of A, if $\bigvee^A M$ exists, then $\bigvee^B M$ also exists and they are equal. (This is the definition of regularity from [34].)

(4) Every maximal disjoint family in A is a maximal disjoint family in B.

---
[5] The countable chain condition, i.e. A contains no uncountable set of pairwise disjoint elements

(5) *For each non-zero $b \in B$ there is some non-zero $a \in A$ such that*
$A \upharpoonright (a - b) = \{0\}$.

(6) *There is a dense subset M of B such that for each non-zero $m \in M$ there*
is some non-zero $a \in A$ such that $A \upharpoonright (a - m) = \{0\}$.

Although these forms are widely known, we include the **proof** for completeness
sake. The implications (1) $\Rightarrow$ (2), (3) $\Rightarrow$ (4), and (5) $\Rightarrow$ (6) are obvious. We
establish the missing links. To distinguish between disjoint complements in A
and B, we use δ and $\vartriangle$, respectively.

(2) $\Rightarrow$ (3). Let $M \subseteq A$ be given such that $a = \bigvee^A M$ exists. Assuming $m \leq b \in$
B for all $m \in M$, we must show $a \leq b$. From $M \subseteq A \upharpoonright b$ and the regularity of
$A \upharpoonright b$, we get

$$M^{\delta\delta} \subseteq (A \upharpoonright b)^{\delta\delta} = A \upharpoonright b.$$

But, $a = \bigvee^A M$ implies $a \in M^{\delta\delta}$. So, $a \in A \upharpoonright b$, i.e. $a \leq b$, as desired.

(4) $\Rightarrow$ (5). Let a non-zero $b \in B$ be given. Choose maximal disjoint subsets M_+
and M_- of $A \upharpoonright b$ and $A \upharpoonright - b$, respectively.

Case 1. There is a non-zero $a \in A$ which is disjoint from all $m \in M_+ \cup M_-$.
To verify $A \upharpoonright (a - b) = \{0\}$ consider any $x \in A$ such that $x \leq a$ and $x \wedge b = 0$. If
x were non-zero, then, belonging to $A \upharpoonright - b$, it would have a non-zero intersection
with some $m \in M_-$. But this is impossible, for $m \wedge x \leq m \wedge a = 0$, by the choice
of a.

Case 2. M is maximal disjoint in A, hence, by (4), also in B. Consequently,
b intersects some $m \in M_+ \cup M_-$. As b is disjoint from all elements of M_-, we
conclude that M_+ must contain a non-zero m. But then $A \upharpoonright (m-b) = A \upharpoonright 0 = \{0\}$,
as desired.

(6) $\Rightarrow$ (1): Let I be regular in B. We have to show $(I \cap A)^{\delta\delta} \subseteq I$. Assuming
that this were false, we take some $c \in (I \cap A)^{\delta\delta}$ not in I. As $I = I^{\vartriangle\vartriangle}$, there is
some $b \in I^{\vartriangle}$ such that $b \wedge c \neq 0$. Applying (6), we first find $0 < m \in M$ such
that $m \leq c \wedge b$ and then $0 < a \in A$ such that $A \upharpoonright (a - m) = \{0\}$.

To check $a \in (I \cap A)^{\delta}$, we take any $d \in I \cap A$. From $d \in I$ and $b \in I^{\vartriangle}$ we get
$0 = d \wedge b \geq d \wedge m$, which together with $a \geq a \wedge d \in A$ yields $a \wedge d \in A \upharpoonright (a-m) =$
$\{0\}$, as desired. As $c \in (I \cap A)^{\delta\delta}$, we must have $c \wedge a = 0$.
On the other hand, $c \wedge a \geq m \wedge a > 0$, for, otherwise, $0 < m \in A \upharpoonright (m-a) = \{0\}$.
$\square$

As principal ideals are regular, we get from (2) and 1.2.1

Corollary 1.2.17 *If $A \leq_{rc} B$, then $A \leq_{reg} B$.*

Now we go over to the next class of embeddings, which is characterized by the
preservation of countably generated ideals.

Definition 1.2.18 $A \leq B$ *is called a σ-embedding, symbolically $A \leq_\sigma B$, if*
$A \upharpoonright b$ *is countably generated for all $b \in B$.* $\square$

For the sake of completeness we also mention dense embeddings. In the above style they can be defined by demanding that the property $I \neq \{0\}$ be preserved. We don't waste good ink to write down transitivity properties of regular, dense and σ-embeddings. They are obvious with the possible exception of

Lemma 1.2.19 *If $A \leq B \leq C$, where $A \leq_{reg} C$ and B is finitely generated over A, then $B \leq_{reg} C$.*

Proof. As in the proof of 1.2.7, we can assume $B = A(x)$. Let $c \in C \setminus \{0\}$ be given. Then either $c \wedge x$ or $c - x$ is non-zero. We confine ourselves to the first case (otherwise replace x by $-x$). Use $A \leq_{reg} C$ to choose $a \in A \setminus \{0\}$ such that $A \restriction (a - (c \wedge x)) = \{0\}$. Then $0 < a \wedge (c \wedge x) \leq a \wedge x \in B$ and it will be sufficient to establish that $B \restriction ((a \wedge x) - (c \wedge x)) = \{0\}$. Given that any $b \in B$ can be written as $b = (a_1 \wedge x) \vee (a_2 - x)$ with $a_1, a_2 \in A$, any $b \in B \restriction (((a \wedge x) - (c \wedge x))$ would be of the form $b = a_1 \wedge x$. Then $a \wedge a_1 \in A \restriction (a - (c \wedge x)) = \{0\}$, hence $b = a_1 \wedge x = (a_1 \wedge x) \wedge (a \wedge x) = 0$, as desired. $\square$

We end this section by some more trivial but useful

Observations 1.2.20

(1) *By 1.2.16(5), dense embeddings are regular.*

(2) *If $A \leq_{rc} B$, then $A \leq_{\sigma} B$.*

(3) *If $A \leq_{reg} B$ and all regular ideals of A are countably generated, then $A \leq_{\sigma} B$.*

The next observation will be used several times

Observation 1.2.21 *Assume $A \leq B$ and let I be an ideal of B maximal with respect to the property that $A \cap I = \{0\}$. Then the canonical projection $a \mapsto a/I$ injectively maps A onto a dense subalgebra of B/I.*
It follows that the ideal $A \restriction (b \vee I) = \{a \in A : a \leq b \vee i$ for some $i \in I\}$ is regular for all $b \in B$

Indeed, $A \cap I = \{0\}$ guaratees injectivity. To check density, let $0 < b/I \in B/I$ be given. Then $b \notin I$, hence, by maximality, there is a non-zero $a \in A$ in the ideal generated by $I \cup \{b\}$. In other words, $0 < a \leq i \vee b$ for some $i \in I$. It follows that $a/I \leq b/I$ and $a/I > 0$, by injectivity.

The second assertion follows from the first and 1.2.20(1). $\square$

Topological duality. Every embedding $A \leq B$ corresponds to a surjective continuous mapping $Ult\, B \longrightarrow Ult\, A$, sending U to $U \cap A$. If $A \leq_{rc} B$, then the dual mapping is open, i.e. images of open sets are open. Regular embeddings correspond to semi-open or skeletal mappings, i.e. images of non-empty open sets have non-empty interior. The concepts corresponding to dense and σ-embeddings are irreducible and functionally closed mappings, respectively.

1.3 Characterizations of projective Boolean algebras

Although the material of this section is well-known, we present it with proofs (mostly taken from [35]). This is done for completeness sake and because these proofs should be compared with later proofs in more general situations.

Definition 1.3.1 A Boolean algebra A is projective iff

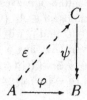

i.e. iff for all homomorphisms $A \xrightarrow{\varphi} B \xleftarrow{\psi} C$, with ψ surjective, there exists a homomorphism $\varepsilon : A \to C$ such that $\psi \circ \varepsilon = \varphi$. $\square$

There are a lot of alternative characterizations of projectivity, many of them listed in the following theorem. The expressions 'rc-skeleton' and 'rc-filtration' occuring in it are shorthand for 'skeleton (resp. filtration) whose elements are relatively complete subalgebras'. In the same way we shall use the expressions 'regular skeleton', 'σ-filtration', etc. These 'nice collections of well embedded subalgebras' will accompany us throughout.

Theorem 1.3.2 *The following assertions are equivalent for all Boolean algebras* A.

(1) A *is projective.*

(2) A *is a retract of a free Boolean algebra.*

(3) (Širokov [56]) *There are a free algebra* F, *a homomorphism* $\varphi : F \to A$ *and a mapping* $\varepsilon : A \to F$ *such that* $\varphi \circ \varepsilon = id_A$ *and* ε *preserves* 0 *and* $\wedge$.

(4) (Ščepin [55]) A *has an additive rc-skeleton.*

(5) *There exists a family* $(A_j)_{j \in J}$ *of countable subalgebras of* A *such that* $A = \bigcup_{j \in J} A_j$ *and* $\left(\bigcup_{j \in K} A_j\right) \leq_{rc} A$, *for all* $K \subseteq J$.

(6) (Haydon [23]) A *has a well-ordered rc-filtration* $(A_\alpha)_{\alpha < \lambda}$ *such that* A_0 *is countable and each* $A_{\alpha+1}$ *is countably generated over* A_α.

(7) (Koppelberg [35]) A *has a well-ordered rc-filtration* $(A_\alpha)_{\alpha < \lambda}$ *starting with* $A_0 = 2$ *and such that each* $A_{\alpha+1}$ *is generated by* $A_\alpha \cup \{a_\alpha\}$ *for a suitable element* a_α.

Proof. (1) $\Rightarrow$ (2). Let $\psi : F \to A$ be any surjective homomorphism, where F is free and of suitable cardinality. Then, by projectivity, for $A \xrightarrow{id} A \xleftarrow{\psi} F$, there exists $\varepsilon : A \to F$ such that $\psi \circ \varepsilon = id$. So A is a retract of F.

As (2) $\Rightarrow$ (3) is trivial we go over to (3) $\Rightarrow$ (4). Let F be freely generated by the set U, and let $\varphi : F \to A$ and $\varepsilon : A \to F$ be as in (3). We call a subset V of U *admissible* if for all $v \in V$, $\varepsilon(\varphi(v))$ and $\varepsilon(\varphi(-v))$ belong to $\langle V \rangle$. It is not hard to prove that

CLAIM 1. *Each subset of U is contained in an admissible subset of the same power.*

and

CLAIM 2. *The union of an arbitrary collection of admissible sets is admissible.*

From claim 1 we easily get that

$$S = \{\langle \varphi(V) \rangle : V \subseteq U \text{ is admissible}\}$$

is absorbing. Claim 2 yields that S is an additive skeleton. Indeed,

$$\langle \bigcup_{i \in I} \langle \varphi(V_i) \rangle \rangle = \langle \bigcup_{i \in I} \varphi(V_i) \rangle = \langle \varphi(\bigcup_{i \in I} V_i) \rangle.$$

It remains to see that S consists of relatively complete subalgebras. Take an admissible V and an arbitrary $a \in A$. (A look at the diagram below may be helpful.)

$$
\begin{array}{ccc}
\langle V \rangle & \leq_{rc} & F \\
\varphi \Big\downarrow \Big\uparrow \varepsilon & & \varphi \Big\downarrow \Big\uparrow \varepsilon \\
\langle \varphi(V) \rangle & \leq & A
\end{array}
$$

It is sufficient to find a minimal element of $\varphi(\langle V \rangle)$ above a. By 1.2.5, $\langle V \rangle \leq_{rc} F$. So we can take a minimal element $c \in \langle V \rangle$ above $\varepsilon(a)$. Then $\varphi(c) \geq \varphi(\varepsilon(a)) = a$. If $\varphi(c)$ were not minimal above a, then we could find some $d \in \langle \varphi(V) \rangle$ such that $d \wedge a = 0$ and $d \wedge \varphi(c) \neq 0$. Without loss, we can assume that d is of the form

$$d = \varphi(v_1) \wedge \ldots \wedge \varphi(v_n) \wedge \varphi(-v_{n+1}) \ldots \wedge \varphi(-v_m)$$

As ε preserves $\wedge$ and V is admissible, we would have $\varepsilon(d) \in \langle V \rangle$ and

$$0 = \varepsilon(0) = \varepsilon(d \wedge a) = \varepsilon(d) \wedge \varepsilon(a).$$

But then $\varepsilon(d) \wedge c = 0$, because c was minimal above $\varepsilon(a)$. It follows that

$$d \wedge \varphi(c) = \varphi(\varepsilon(d)) \wedge \varphi(c) = \varphi(\varepsilon(d) \wedge c) = 0,$$

which is the desired contradiction.

(4) $\Rightarrow$ (5). Let $\mathcal{S}$ be as in (4) and let $(A_j)_{j \in J}$ enumerate $\{S \in \mathcal{S} : |S| \leq \aleph_0\}$.

(5) $\Rightarrow$ (6). Without loss we can assume that the family existing by (5) is indexed by an ordinal: $(B_\alpha)_{\alpha < \lambda}$. It is then sufficient to put $A_\alpha = \langle \bigcup_{\beta < \alpha} B_\beta \rangle$ to arrive at a filtration as demanded in (6).

(6) $\Rightarrow$ (7). Let $(A_\alpha)_{\alpha < \lambda}$ be as in (6). There is no loss in assuming $A_0 = 2$, for we can simply put the two-element algebra in front if it is not there already.

We now modify this filtration in order to achieve that all successor steps are realized by one-element extensions.

The idea is to insert additional subalgebras in between the given A_α and $A_{\alpha+1}$. Let $(a_n)_{n < m}$ enumerate a countable set (so $m \leq \omega$) such that $A_{\alpha+1} = A_\alpha(\{a_n : n < m\})$ and refine

$$A_\alpha \leq A_\alpha(a_0) \leq A_\alpha(a_0, a_1) \leq \ldots \leq A_\alpha(a_0, a_1, \ldots, a_n) \leq \ldots \ldots A_{\alpha+1}.$$

All the newly inserted algebras are finitely generated over relatively complete subalgebras of A and, therefore, also relatively complete in A (by 1.2.7).

(7) $\Rightarrow$ (1). Let $(A_\alpha)_{\alpha < \lambda}$ be a filtration as in (7) and consider homomorphisms $A \xrightarrow{\varphi} B \xleftarrow{\psi} C$, with ψ surjective. By induction on α we construct homomorphisms $\varepsilon_\alpha : A_\alpha \to C$ such that $\varepsilon_\alpha \restriction A_\beta = \varepsilon_\beta$ for $\beta < \alpha$ and $\varphi \restriction A_\alpha = \psi \circ \varepsilon_\alpha$ for all α. This will end the proof, for the desired ε then simply is $\bigcup_{\alpha < \lambda} \varepsilon_\alpha$.

As $A_0 = 2$, there is no choice and no problem with ε_0. At limit steps there is no problem either; as the filtration is continuous, we may put $\varepsilon_\beta = \bigcup_{\alpha < \beta} \varepsilon_\alpha$.

So let us consider the successor step assuming that ε_α exists already. The problem is illustrated by the diagram below.

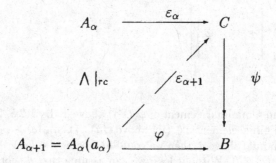

Using $A_\alpha \leq_{rc} A_{\alpha+1} = A_\alpha(a_\alpha)$ we may take $a_- \leq a_\alpha \leq a_+$ where a_- and a_+ are maximal and minimal in A_α, respectively. Clearly $\varepsilon_\alpha(a_-) \leq \varepsilon_\alpha(a_+)$.

By assumption, ψ is surjective. Therefore, we may take some $x \in C$ such that $\psi(x) = \varphi(a_\alpha)$. Let us put

$$y = (\varepsilon_\alpha(a_-) \vee x) \wedge \varepsilon_\alpha(a_+) = \varepsilon_\alpha(a_-) \vee (x \wedge \varepsilon_\alpha(a_+))$$

and show that there is a homomorphism $\varepsilon_{\alpha+1} : A_{\alpha+1} = \langle A_\alpha \cup \{a_\alpha\} \rangle \longrightarrow C$ extending $\varepsilon_\alpha \cup \{(a_\alpha, y)\}$.

We use Sikorski's Extension Criterion 0.0.1, which in our case boils down to the following two claims holding for each $a \in A_\alpha$.

$$(i) \quad a \wedge a_\alpha = 0 \quad \Rightarrow \quad \varepsilon_\alpha(a) \wedge y = 0, \quad \text{and}$$
$$(ii) \quad a \leq a_\alpha \quad \Rightarrow \quad \varepsilon_\alpha(a) \leq y.$$

To check (i) we notice that, by the minimality of a_+, $a \wedge a_\alpha = 0$ implies $a \wedge a_+ = 0$, hence $\varepsilon_\alpha(a) \wedge \varepsilon_\alpha(a_+) = 0$. From this the claim follows because $y \leq \varepsilon_\alpha(a_+)$. Claim (ii) is established in the same way using the maximality of a_- below a_α.

So, $\varepsilon_{\alpha+1}$ exists. By construction, it extends ε_α. From

$$\psi(\varepsilon_{\alpha+1}(a_\alpha)) = \psi(y) = \psi((\varepsilon_\alpha(a_-)) \vee \psi(x)) \wedge \psi(\varepsilon_\alpha(a_+))$$

$$= (\varphi(a_-) \vee \varphi(a_\alpha)) \wedge \varphi(a_+) = \varphi((a_- \vee a_\alpha) \wedge a_+) = \varphi(a_\alpha)$$

and $\psi \circ \varepsilon_\alpha = \varphi | A_\alpha$ it follows that

$$\psi \circ \varepsilon_{\alpha+1} = \varphi$$

is true on a set of generators of $A_{\alpha+1}$, hence on all of it. $\square$

Topological duality: The dual spaces of projective Boolean algebras are so-called absolute extensors in dimension zero, AE(0), and also known as Dugundji spaces.

1.4 Free extensions, Sirota's Lemma and Ščepin's Theorem

In this section we meet another aproach to the classification of embeddings $A \leq B$. In contrast to the previous types, which were concerned with the position of A inside B, we are now interested in the position of B above A. Therefore, it would be more appropriate to speak about classes of extensions. For a fixed Boolean algebra A, we consider the category $\mathcal{B}_A$ consisting of all Boolean algebras that extend A. Morphisms of $\mathcal{B}_A$ are those usual homomorphisms that keep A pointwise fixed. Now categorical properties of $B \in \mathcal{B}_A$ can be interpreted as properties of $A \leq B$. In this section we study one such property, namely freedom. We do this, however, in a more technical way, which does not mention the category $\mathcal{B}_A$ explicitly.

Recall that two subalgebras $A, B \leq C$ are *independent* (we also say that A is independent of B), provided that $a \wedge b \neq 0$ for all non-zero $a \in A$ and $b \in B$.

It is an easy consequence of Sikorski's Extension Criterion 0.0.1 that $C \simeq A \otimes B$ if C is generated by its independent subalgebras A and B (cf. 11.4 in [34]).

An element x is said to be independent of A if $\langle x \rangle$ and A are independent in the above sense, i.e. if $a \wedge x \neq 0$ and $a - x \neq 0$ for all non-zero $a \in A$.

Definition 1.4.1 We say that B is a *free extension* of A and write $A \leq_{free} B$ iff there exists an infinite free subalgebra $F \leq B$ which is independent of A and such that $B = \langle A \cup F \rangle$. $\square$

In view of the above remark, $A \leq_{free} B$ iff $B = A \otimes \mathrm{Fr}\,\kappa$ for some infinite κ. More precisely, $A \leq_{free} B$ iff the embedding $A \leq B$ is isomorphic to the canonical embedding $A \leq A \otimes \mathrm{Fr}\,\kappa$ (identifying $a \in A$ with $a \otimes 1 \in A \otimes \mathrm{Fr}\,\kappa$). We start the consideration of free extension with some easy

Observations 1.4.2

(1) *Clearly,* $2 \leq_{free} A$ *iff* A *is free.*

(2) *If* $A \leq_{free} B$ *and* $A' \leq B$ *is finitely generated over* A, *then* $A' \leq A'' \leq_{free} B$ *for some* A'' *which is also finitely generated over* A.

(3) $A \leq_{free} B$ *implies* $A \leq_{rc} B$, *by 1.2.5.*

As to transitivity, we have

Lemma 1.4.3 (1) *If* $A \leq_{free} B$ *and* $B \leq_{free} C$, *then* $A \leq_{free} B$.

(2) *Let* $(A_\alpha)_{\alpha < \lambda}$ *be a well-ordered filtration of some Boolean algebra* A. *If* $A_\alpha \leq_{free} A_{\alpha+1}$ *for all* $\alpha < \lambda$, *then* $A_\beta \leq_{free} A$ *for all* $\beta < \lambda$.

Proof. (1) Let F and G witness $A \leq_{free} B$ and $B \leq_{free} C$, respectively. We check that $H = \langle F \cup G \rangle$ witnesses $A \leq_{free} C$. Clearly,

$$\langle A \cup H \rangle = \langle A \cup \langle F \cup G \rangle \rangle = \langle \langle A \cup F \rangle \cup G \rangle = \langle B \cup G \rangle = C.$$

To see that A and H are independent, take non-zero $a \in A$ and $z \in H$. Then $z = \bigvee_{i=1}^{n} x_i \wedge y_i$ for suitable $x_i \in F$ and $y_i \in G$. Without loss, $x_1 \wedge y_1 \neq 0$. The independence of A and F yields $a \wedge x_1 \neq 0$. As $a \wedge x_1 \in \langle A \cup F \rangle = B$, which is independent of G, we get $(a \wedge x_1) \wedge y_1 \neq 0$, hence $a \wedge z \neq 0$, as desired.

A very similar argument (see the proof of (2) below) shows that $X \cup Y$ freely generates H, provided that X and Y freely generate F and G, respectively. It follows that H is free, which ends the proof of (1).

To prove (2) we fix F_α witnessing $A_\alpha \leq_{free} A_{\alpha+1}$ and let $\beta < \lambda$ be given. We show that

$$F = \langle \bigcup \{ F_\alpha : \beta \leq \alpha < \lambda \} \rangle$$

witnesses $A_\beta \leq_{free} A$. An easy induction shows that $A_\alpha \leq \langle A_\beta \cup F \rangle$ for all α, hence $\langle A_\beta \cup F \rangle = A$.

To see that F is free, we take sets X_α of free generators of F_α. Then, clearly,

$$X = \bigcup \{ X_\alpha : \beta \leq \alpha < \lambda \}$$

generates F. To see that X consists of free generators, we have to check that all intersections

$$(*) \quad x_1 \wedge x_2 \wedge \ldots \wedge x_n$$

are non-zero, where the x_i belong to $X \cup -X$ and no element occurs together with its complement. Sorting the x_i according to the X_α they belong to and intersecting the elements of each group, we transform $(*)$ into an intersection

$$y_1 \wedge y_2 \wedge \ldots \wedge y_k$$

such that $y_i \in F_{\alpha_i}$ with $\alpha_1 < \alpha_2 < \ldots < \alpha_k$. As the X_α consist of *free* generators of F_α, all y_i are non-zero.

We now use induction on $i \leq k$ to check that

$$y_1 \wedge y_2 \wedge \ldots \wedge y_i \neq 0.$$

his is clear for $i = 1$. In the induction step we notice that

$$y_1 \wedge y_2 \wedge \ldots \wedge y_i \quad \text{belongs to} \quad \langle F_{\alpha_1} \cup F_{\alpha_2} \cup \ldots \cup F_{\alpha_i}\rangle \leq A_{\alpha_i + 1} \leq A_{\alpha_{i+1}}.$$

As $y_{i+1} \in F_{\alpha_{i+1}}$, which is independent of $A_{\alpha_{i+1}}$, we get

$$(y_1 \wedge y_2 \wedge \ldots y_i) \wedge y_{i+1} \neq 0,$$

as desired. The freedom of F is established. In a similar way (cf. the argument in the proof of (1)) one checks that F is independent of A_β. $\square$

Corollary 1.4.4 *Let* $(A_\alpha)_{\alpha \leq \gamma}$ *be a well-ordered filtration of* A_γ.

(1) *If* $A_\alpha \leq_{free} A_{\alpha+1}$ *for all* $\alpha < \gamma$, *then* $A_\alpha \leq_{free} A_\beta$ *for all* $\alpha < \beta \leq \gamma$.

(2) *If, in addition,* A_0 *is free, then all* A_α *are free.*

Indeed, (1) follows by an easy induction from the above lemma. Assertion (2) is an immediate consequence of (1) and observation 1.4.2(1). $\square$

Below we need the following

Lemma 1.4.5 *If* $A \leq_{free} B$ *and* A *satisfies the ccc, then so does* B.

Proof. Let $F \subseteq B$ witness $A \leq_{free} B$. Assume by contradiction, that there were a family $(b_\alpha)_{\alpha < \omega_1}$ of pairwise disjoint elements of B. Shrinking them a little, each b_α can be written in the form

$$b_\alpha = a_\alpha \wedge f_\alpha,$$

where $a_\alpha \in A$ and $f_\alpha \in F$. By a well-known property of free Boolean algebras[6], there is a subset $W \subseteq \omega_1$ of cardinality $\aleph_1$ such that $f_\alpha \wedge f_\beta \neq 0$ for all $\alpha, \beta \in W$. Take $\alpha, \beta \in W$. From

$$0 = b_\alpha \wedge b_\beta = \underbrace{(a_\alpha \wedge a_\beta)}_{\in A} \wedge \underbrace{(f_\alpha \wedge f_\beta)}_{\neq 0;\ \in F}$$

---

[6] In 2.7.2 below we prove a stronger assertion for a wider class of algebras.

and the independence of A and F it follows that $a_\alpha \wedge a_\beta = 0$. So $(a_\alpha)_{\alpha \in W}$ is an uncountable disjoint family in A, which contradicts the ccc holding in A. $\square$

To close this general part about free extensions we introduce the *standard filtration* of an uncountable free Boolean algebra, F say. This is a well-ordered filtration $(F_\alpha)_{\alpha < \rho}$ of F, where ρ is the first ordinal of cardinality $|F|$ such that

(*i*) F_0 is countably infinite and free,

(*ii*) $F_\alpha \leq_{free} F_{\alpha+1}$ for all α, and

(*iii*) each $F_{\alpha+1}$ is countably generated over F_α.

By the above corollary, all F_α are free and F freely extends each of them.

It should be clear that each uncountable free Boolean algebra has a standard filtration. Just decompose the set of free generators, X say, into a continuous chain of subsets $X = \bigcup_{\alpha < \rho} X_\alpha$ such that X_0 and all differences $X_{\alpha+1} \setminus X_\alpha$ are countably infinite. Then $F_\alpha = \langle X_\alpha \rangle$ is as desired.

Moreover, it is not hard to prove (exercise) that any two standard filtrations of F are isomorphic. That is why we speak of *the* standard filtration.

Our second aim in this section is to characterize *countably generated* free extensions. In this characterization the following notation will be convenient.

Definition 1.4.6 We write $A \leq_s B$ to express that each ultrafilter $U \in Ult\,A$ *splits* over B, i.e. has at least two different extensions to ultrafilters of B. We write $A \leq_{sp} B$ if each ultrafilter of A *splits perfectly* over B, i.e. $A' \leq_s B$ holds for all $A' \leq B$ that are finitely generated over A. $\square$

Here are a couple of more or less obvious

Observations 1.4.7

(1) *If $A \leq B \leq C$ and either $A \leq_s B$ or $B \leq_s C$, then $A \leq_s C$.*

(2) *If $A \leq B \leq C$ and $B \leq_{sp} C$, then $A \leq_{sp} C$.*

(3) *If B contains an element which is independent of A, then $A \leq_s B$.*

(4) *If $(A_n)_{n < \omega}$ is a chain and $A_n \leq_s A_{n+1}$ for all n, then $A_n \leq_{sp} \bigcup_{m < \omega} A_m$.*

(5) *A is atomless iff $A' \leq_{sp} A$ for all finite $A' \leq A$.*

(6) *More generally, $A \leq_{sp} B$ iff $B/\hat{U}$ is atomless for each ultrafilter U of A, where $\hat{U}$ denotes the filter generated by U in B.*

From 1.4.2(2) and 1.4.7(3) we immediately get

Observation 1.4.8 $A \leq_{free} B$ *implies* $A \leq_{sp} B$.

Now we come to more serious matters.

Lemma 1.4.9 *If $A \leq_{rc} B$ and $A \leq_s B$, then for each $b \in B$ there is some $x \in B$, which is independent of A and such that $b \in A(x)$.*

Proof. We first forget about b and show that there is some $y \in B$, which is independent of A, i.e. such that

$$a \wedge y \neq 0 \quad \text{and} \quad a - y \neq 0$$

for all non-zero $a \in A$. Let $q : B \to A$ denote the upper projection and notice that the independence of y is equivalent to

$$q(-y) = 1 \quad \text{and} \quad q(y) = 1.$$

In other words, we have to show that 1 belongs to the ideal (by 1.2.10)

$$I = \{q(z) \wedge q(-z) : z \in B\} \subseteq A.$$

Assuming that this is false we can extend I to a prime ideal of A and consider the dual ultrafilter U, say. By definition,

$$-q(z) \vee -q(-z) \in U, \text{ for all } z \in B.$$

A contradiction with $A \leq_s B$ will be reached if we show that the filter $\hat{U}$ generated by U in B is in fact an ultrafilter, i.e. includes one of $z, -z$ for all $z \in B$.

Let z be given. As $-q(z) \vee -q(-z)$ belongs to the *ultra*filter U, either $-q(z)$ or $-q(-z)$ belongs to U. In the first case, $-q(z) \leq -z \in \hat{U}$. In the second case $-q(-z) \leq z \in \hat{U}$, as desired.

Half of the work is done; we have got $y \in B$ independent of A. Now, remembering about the given b, we take this y and put

$$x = [b \wedge (q(b) \wedge q(-b))] \vee [y - (q(b) \wedge q(-b))].$$

From $-q(-b) \leq b \leq q(b)$ we get

$$b - q(-b) = -q(-b) \text{ and } b \wedge q(-b) = b \wedge q(b) \wedge q(-b) = x \wedge (q(b) \wedge q(-b)),$$

hence

$$b = [b \wedge q(-b)] \vee [b - q(-b)] = [x \wedge (q(b) \wedge q(-b))] \vee -q(-b) \in A(x).$$

As noticed above, the independence of x and A is equivalent to $q(x) \wedge q(-x) = 1$. The calculations proving this equation become easier if we introduce the notation $i(z) = q(z) \wedge q(-z)$ and use the formula $i(a \wedge z) = a \wedge i(z)$ which was established for $a \in A$ and $z \in B$ in the proof of 1.2.10. The element x can be rewritten as $x = [b \wedge i(b)] \vee [y - i(b)]$. As $i(b) \in A$ we get

$$i(b) \wedge i(x) = i(i(b) \wedge x) = i(i(b) \wedge b) = i(b) \wedge i(b) = i(b)$$

and

$$-i(b) \wedge i(x) = i(-i(b) \wedge x) = i(y - i(b)) = -i(b) \wedge i(y) = -i(b) \wedge 1 = -i(b),$$

which together show $i(x) = 1$. $\square$

Theorem 1.4.10 (Sirota's Lemma [57]) *Assume that B is countably generated over A. Then $A \leq_{free} B$ iff $A \leq_{rc} B$ and $A \leq_{sp} B$.*

Proof. We have seen above that both conditions are necessary (even without the countability assumption). To prove that they are sufficient, we choose a sequence $(b_n)_{n<\omega}$ such that $B = \langle A \cup \{b_n : n < \omega\}\rangle$. By induction we then find elements $x_n \in B$ such that

    (i) $b_n \in A(x_0, \ldots x_n)$, and   (ii) x_n is independent of $A(x_0, \ldots, x_{n-1})$.

After we have got these elements, it remains to see that $F = \langle\{x_n : n < \omega\}\rangle$ witnesses $A \leq_{free} B$. From (i) it follows that $B = \langle A \cup F\rangle$. Using (ii) the independence of A and F is easily proved as in 1.4.3 above. In each step of the construction we are practically in the same situation:

$$A(x_0 \ldots, x_n) \leq_{rc} B \quad \text{and} \quad A(x_0 \ldots, x_n) \leq_{sp} B,$$

by 1.2.7(3) and the very definition of $\leq_{sp}$. Therefore, the above lemma allows us to find x_{n+1}. $\square$

The following result, is one of the most beautiful and useful of the whole theory. It is due to Ščepin (cf. also 3.4 in [35]) and generalizes the well-known fact that all countable atomless Boolean algebras are free.

Theorem 1.4.11 (Theorem 9 in [53]) *Let A be a projective Boolean algebra of uncountable cardinality. If no ultrafilter of A is generated by less than $|A|$ elements, then A is free.*

Proof. Let S be an additive rc-skeleton of A, which exists by assertion (4) of the theorem 1.3.2. There is no loss in assuming $2 \in S$. The main work will be in establishing the following

CLAIM *For each $S \in \mathcal{S}$ and $a \in A$, if $|S| < |A|$, then $a \in T$ for some $T \in \mathcal{S}$ such that $S \leq_{free} T$ and T is countably generated over S.*

Let S be given. First we take $S \leq S_0 \in \mathcal{S}$ such that $a \in S_0$. By additivity, it is possible to choose S_0 countably generated over S. Assume that we have constructed $S \leq S_0 \leq_s S_1 \leq_s \ldots \leq_s S_n$, with all S_i countably generated over S and in $\mathcal{S}$. Then $|S_n| = |S| < |A|$ and, by assumption, no ultrafilter of A can be generated by its intersection with S_n. In other words, $S_n \leq_s A$. Moreover, $S_n \in \mathcal{S}$ implies $S_n \leq_{rc} A$. It follows that we can use lemma 1.4.9 to choose an element of A which is independent of S_n. This element can be captured in some $S_{n+1} \in \mathcal{S}$, which is countably generated over S_n. The independent element guarantees $S_n \leq_s S_{n+1}$, by 1.4.7(3).

Put $T = \bigcup_{n<\omega} S_n \in \mathcal{S}$. Then $S \leq_{sp} T$, by 1.4.7(4). Moreover, $S \leq_{rc} A$ implies $S \leq_{rc} T$. As T is also countably generated over S, we get $S \leq_{free} T$, by Sirota's Lemma. The claim is proved.

Let ρ denote the first ordinal of power $|A|$. Using the claim at successor steps it is easy to produce a continuous filtration $(S_\alpha)_{\alpha<\rho}$ of A such that (*i*) S_0 is countably infinite and free (privately start the construction with $S_{-1} = 2$),

(*ii*) $S_\alpha \leq_{free} S_{\alpha+1}$ for all α

(*iii*) each $S_{\alpha+1}$ is countably generated over S_α, and, to keep the induction going,

(*iv*) all S_α belong to $\mathcal{S}$.

By 1.4.4(2), this filtration witnesses that A is free. $\square$

Let us mention the following nice

Corollary 1.4.12 *The following assertions are equivalent for an arbitrary Boolean algebra* A.

(1) *A is projective.*

(2) *There is a free Boolean algebra such that $A \otimes F$ is free.*

(3) *There is some Boolean algebra B such that $A \otimes B$ is free.*

Proof. Notice first that the class of projective Boolean algebras is closed under free products. Indeed, if A_1 and A_2 are retracts of the free algebras F_1 and F_2, then $A_1 \otimes A_2$ is easily seen to be a retract of $F_1 \otimes F_2$, the latter remaining free.

Now, if A is projective, then $A \otimes F$ is projective for all free F. Taking F big enough, it overrides A in the sense that no ultrafilter of $A \otimes F$ is generated by less that $|F| = |A \otimes F|$ elements. So, $A \otimes F$ is free, by Ščepin's theorem.

The implication (2) $\Rightarrow$ (3) is trivial and (3) $\Rightarrow$ (1) is due to the fact that A is a retract of $A \otimes B$. $\square$

1.5 Projective embeddings

Definition 1.5.1 We call an embedding $A \leq B$ *projective* and write $A \leq_{proj} B$ if for all Boolean algebras C and all pairs of homomorphisms $A \xrightarrow{\iota} C \xrightarrow{\pi} B$ such that $\pi \circ \iota = id_A$ and π is surjective there is a homomorphism $\varepsilon : B \longrightarrow C$ extending ι such that $\pi \circ \varepsilon = id_B$. $\square$

The following diagram illustrates this definition.

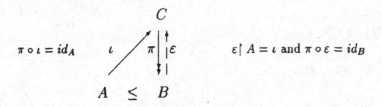

$$\pi \circ \iota = id_A \qquad \iota \quad \diagup \quad \pi \Big| \Big| \varepsilon \qquad \varepsilon \restriction A = \iota \text{ and } \pi \circ \varepsilon = id_B$$

$$A \quad \leq \quad B$$

As this notion is rather categorical, it would be more appropriate to speak about projective extensions. $A \leq_{proj} B$ can be defined by saying that B is a projective object in the category $\mathcal{B}_A$ defined at the beginning of the previous section. Our definition above is more technical, which makes it better suited for applications. The interested reader may prove the equivalence of both definitions as an exercise.

The proof that the classes of projective algebras and embeddings are adequate will be based on the following characterizations that will also be useful elsewhere. In assertion (3) we need a relativized version of the notion of skeleton. We say that S is a *skeleton of B over A* iff S is a collection of subalgebras of B such that

(*i*) all $S \in S$ contain A,

(*ii*) S is closed under unions of chains, and

(*iii*) for each $C \leq B$ such that $A \leq C$ there exists $S \in S$ such that $C \leq S$ and $w(S/A) \leq w(C/A)$

where, for an arbitrary embedding $C \leq E$, the cardinal number $w(E/D)$ is defined as $\min\{|X| : X \subseteq E; \langle D \cup X \rangle = E\}$ and called the *weight of E over D*.

Theorem 1.5.2 *The following assertions are equivalent for an arbitrary embedding $A \leq B$.*

(1) $A \leq_{proj} B$

(2) *There is a free extension C of A and a pair of homomorphisms $\pi : C \to B$ and $\varepsilon : B \to C$ such that $\varepsilon \circ \pi = id_B$ and $\pi(a) = \varepsilon(a) = a$ for all $a \in A$. (This is the definition of projectivity from [35].)*

(3) *There is an additive rc-skeleton of B over A.*

(4) *B has a well-ordered rc-filtration $(B_\alpha)_{\alpha < \lambda}$ such that $B_0 = A$ and each $B_{\alpha+1}$ is countably generated over B_α.*

Assertion (2) can be illustrated by the diagram below. It says, in other words, that B is a retract of a free object of $\mathcal{B}_A$.

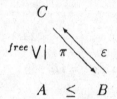

Proof. (1) $\Rightarrow$ (2). Let F be a free Boolean algebra and $\varphi : F \to B$ a surjective homomorphism. Identifying $a \in A$ with $a \otimes 1$ in $A \otimes F$, we can consider $A \otimes F$

as a free extension of A. We define $\pi : A \otimes F \to B$ by $\pi(a \otimes f) = a \wedge \varphi(f)$. Then the desired $\varepsilon : B \to A \otimes F$ comes from the definition of projective embedding.

$(2) \Rightarrow (3)$. Let C, π, and ε be as in (2). Take a free subalgebra $F \leq C$ which witnesses $A \leq_{free} C$. Let $U \subseteq F$ be a set of free generators of F. Define a subset V of U to be admissible if $\varepsilon(\pi(\langle A \cup V \rangle)) \subseteq \langle A \cup V \rangle$. Then proceed as in the proof of $1.3.2(3) \Rightarrow (4)$. Due to the fact that this time ε is a homomorphism, the argument becomes even easier.

$(3) \Rightarrow (4)$ should be obvious. To prove $(4) \Rightarrow (1)$ one first uses the same argument that proved $1.3.2(6) \Rightarrow (7)$ to refine the given filtration in such a way that all the successor steps become simple, i.e. $B_{\alpha+1} = B_\alpha(b_\alpha)$. Then, given any $A \xrightarrow{\iota} C \xrightarrow{\pi} B$, one extends ι by induction to $\varepsilon_\alpha : B_\alpha \to C$ such that $\pi \circ \varepsilon_\alpha = id_{B_\alpha}$. This is illustrated in the diagram below and done as in the proof of $1.3.2(7) \Rightarrow (1)$. □

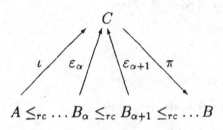

Comparing 1.3.2 and 1.5.2, we immediately get

Corollary 1.5.3 B is projective iff $2 \leq_{proj} B$

Just slightly more complicated is the following

Proposition 1.5.4 *The classes of projective Boolean algebras and projective embeddings are adequate, i.e. the following two assertions are satisfied.*

$(A1)$ *Each projective algebra has a skeleton S such that $S \leq_{proj} T$ for all $S \leq T$ in S.*

$(A2)$ *If B has a filtration $(B_\alpha)_{\alpha < \rho}$ such that all B_α are projective and $B_\alpha \leq_{proj} B_\beta$ for all $\alpha < \beta < \rho$, then B is projective and $B_\alpha \leq_{proj} B$, for all $\alpha < \rho$.*

Proof. (A1) Let B be the given projective algebra. Fix a family $(A_j)_{j \in J}$ as in 1.3.2(5) and put
$$S = \{\langle \textstyle\bigcup_{j \in K} A_j \rangle : K \subseteq J\}.$$
An easy verification shows that S is a skeleton of B.
Let $S = \langle \bigcup_{j \in K} A_j \rangle \leq T = \langle \bigcup_{j \in L} A_j \rangle$ be two elements of S. To see that, indeed, $S \leq_{proj} T$ it is sufficient to verify that

$$\{\langle \textstyle\bigcup_{j \in M} A_j \rangle : K \subseteq M \subseteq K \cup L\}$$

is an additive rc-skeleton of T over S. This is routine.

(A2) is even easier. Given that B_0 is projective, we can fix a Haydon-filtration (shorthand for a filtration as in 1.3.2(6)) of B_0. Using $B_0 \leq_{proj} B_1$ in the form 1.5.2(4), the filtration of B_0 can be prolonged to a Haydon-filtration of B_1. And so on. Finally we get a Haydon-filtration of B, whose existence proves the projectivity of B. Moreover, the tails from B_α to B, prove that $B_\alpha \leq_{proj} B$, again by 1.5.2(4). $\square$

It is worth while to retain the following corolloary from the above proof of (A1).

Corollary 1.5.5 *If S is an additive rc-skeleton of A, then $S \leq_{proj} A$ for all $S \in S$.*

Here are a few more useful observations about projective embeddings.

Proposition 1.5.6 *Assume $A \leq_{proj} B$. Then*

(1) $A \leq_{rc} B$

(2) A *is a retract of B.*

(3) *If A satisfies the ccc, then so does B.*

(4) *If A is projective, then B is also projective.*

(5) *If B is projective, then A is also projective.*

Proof. (1) follows from 1.5.2(4) and transitivity of relative completeness, i.e. 1.2.9. To prove (2), define $\iota : A \to A \otimes B$ by $\iota(a) = a \otimes 1$ and $\pi : A \otimes B \to B$ by $\pi(a \otimes b) = a \wedge b$. Then $\pi \circ \iota = id_A$ and, by definition of projectivity, there exists $\varepsilon : B \to A \otimes B$ extending ι and such that $\pi \circ \varepsilon = id_B$.

Take any homomorphism $\psi : B \to 2 \leq A$ and put $\delta = id_A \otimes \psi : A \otimes B \to A$. Then $\delta \circ \varepsilon : B \to A$ satisfies

$$\delta(\varepsilon(a)) = \delta(\iota(a)) = \delta(a \otimes 1) = a \wedge \psi(1) = a$$

for all $a \in A$.

(3) follows from 1.5.2(2) which implies that B embeds into a free extension C of A. By 1.4.5, C satisfies the ccc.

(4) is proved just as in the argument proving (A2) above and (5) follows from (2) and 1.3.2(2). $\square$

The following nice characterization of projective extensions parallels 1.4.12 and will be useful below.

Proposition 1.5.7 *The extension $A \leq B$ is projective iff there is a free Boolean algebra F such that $A \otimes F$ and $B \otimes F$ are isomorphic over A.*

Proof.   Assume first that $A \leq_{proj} B$ and let $(B_\alpha)_{\alpha < \lambda}$ be as in 1.5.2(4). Fix a well-ordered filtration $(F_\alpha)_{\alpha < \lambda}$ of a suitable free Boolean algebra F such that $F_0 = 2$ and each successor step $F_\alpha \leq F_{\alpha+1}$ is a countably generated free extension. Obviously, $(B_\alpha \otimes F_\alpha)_{\alpha < \lambda}$ is a filtration of $B \otimes F$.

On the other hand, each successor step $B_\alpha \otimes F_\alpha \leq B_{\alpha+1} \otimes F_{\alpha+1}$ is countably generated and relatively complete, by 1.2.6. Moreover, writing $F_{\alpha+1}$ as $F_\alpha \otimes H$, where H is infinite and free, we see that

$$B_\alpha \otimes F_\alpha \leq B_{\alpha+1} \otimes F_\alpha \leq_{sp} (B_{\alpha+1} \otimes F_\alpha) \otimes H = B_{\alpha+1} \otimes F_{\alpha+1}.$$

So, by Sirota's Lemma, each successor step is a free extension, hence $A \simeq A \otimes 2 = B_0 \otimes F_0 \leq_{free} B \otimes F$, by 1.4.3(2). Fixing a free subalgebra, G say, that witnesses this fact, we get $A \otimes G \simeq B \otimes F$, where the isomorphism leaves A pointwise fixed. For cardinality reasons, we must have $|G| = |F|$ and, therefore $F \simeq G$, as desired.

To establish the other direction, let $\varphi : A \otimes F \to B \otimes F$ be an isomorphism such that $\varphi(a \otimes 1) = a \otimes 1$. Fix any homomorphism $\psi : F \to 2$ and define $\delta : B \otimes F \to B$ by putting $\delta(b \otimes f) = b \wedge \psi(f)$.

A quick look at the following commutative diagram

$$
\begin{array}{ccc}
A \otimes F & \xrightarrow{\varphi} & B \otimes F \\
\iota_A \uparrow & & \delta \downarrow\uparrow \iota_B \\
A & \leq & B
\end{array}
$$

where $\iota_A(a) = a \otimes 1$ and $\iota_B(b) = b \otimes 1$ will convince the reader, that the mappings $\pi = \delta \circ \varphi$ and $\varepsilon = \varphi^{-1} \circ \iota_B$ are as required in 1.5.2(2). $\square$

Finally we want to warn the reader. In contrast to what happens for relatively complete and regular embeddings, projectivity is not inhereted downwards, i.e. $A \leq_{proj} B$ and $A \leq C \leq B$ do not imply $A \leq_{proj} C$. A counterexample is easy to give. Just take a non-projective subalgebra C of a projective algebra B. Then $2 \leq_{proj} B$, but $2 \nleq_{proj} C$. In section 3.4 we construct a non-projective C which is even relatively complete in a projective Boolean algebra.

Topological duality: Projective embeddings correspond to so called 0-soft mappings. The weight of the embedding $D \leq E$ equals the weight of the dual continuous mapping $Ult\, E \to Ult\, D$.

Chapter 2

rc-filtered Boolean algebras

In this chapter our first generalization of projectivity will be introduced and
rather systematically studied. The first section contains some technical preparations. After this, the class of rc-filtered Boolean algebras will be defined. Then
we investigate its behaviour with respect to the basic algebraic constructions
and compute their cardinal invariants.

2.1   rc-filtrations

In the first part of this section we develop a little technique that will be useful
for several purposes. For some time we fix a well-ordered rc-filtration $(A_\alpha)_{\alpha < \tau}$
of a Boolean algebra A. We let $q_\alpha : A \to A_\alpha$ denote the projections. Clearly,
$\alpha < \beta < \tau$ implies $q_\alpha(a) \geq q_\beta(a) \geq a$. If β is a limit ordinal, then, by continuity
of the filtration, there is some $\alpha_0 < \beta$ (depending on a) such that $q_\beta(a) \in A_{\alpha_0}$;
hence $q_\alpha(a) = q_\beta(a)$ for all $\alpha_0 < \alpha < \beta$.

Definition 2.1.1 For each $a \in A$ we put

$$d(a) = \{\alpha < \tau : q_\alpha(a) > q_{\alpha+1}(a)\}$$

and call it the *jump set* of a. $\square$

Notice that, although this is not reflected in the notation, $d(a)$ depends on the
filtration.

Lemma 2.1.2 $d(a)$ is finite for all $a \in A$.

Proof. If not, we could pick a strictly increasing sequence $(\alpha_n)_{n < \omega}$ of elements of
$d(a)$. Put $\beta = \sup \alpha_n$. As $\beta \geq \alpha_n + 1$ for all n, we get $q_\beta(a) \leq q_{\alpha_n+1}(a) < q_{\alpha_n}(a)$.

   On the other hand, $q_\beta(a) \in A_\beta = \bigcup_{n < \omega} A_{\alpha_n}$. So, $q_\beta(a) \in A_{\alpha_m}$, for some m,
which implies $q_{\alpha_m}(a) \leq q_\beta(a)$, by the definition of q_{α_m}. $\square$

Lemma 2.1.3 *If $a \wedge b = 0$, then either $q_0(a) \wedge q_0(b) = 0$ or there is some
$\alpha \in d(a) \cap d(b)$ such that $q_{\alpha+1}(a) \wedge q_{\alpha+1}(b) = 0$.*

Proof. First notice that for big α we have $q_\alpha(a) \wedge q_\alpha(b) = a \wedge b = 0$. It follows that there is a least ordinal β such that $q_\beta(a) \wedge q_\beta(b) = 0$. If $\beta = 0$, we are done.

From the continuity of the filtration it follows that β can not be a limit ordinal, for, if it were, $q_\beta(a) = q_\alpha(a)$ and $q_\beta(b) = q_\alpha(b)$ for some $\alpha < \beta$.

Assume then that $\beta = \alpha + 1$, i.e.

$$q_{\alpha+1}(a) \wedge q_{\alpha+1}(b) = 0 \text{ and } q_\alpha(a) \wedge q_\alpha(b) \neq 0.$$

It remains to see that $\alpha \in d(a) \cap d(b)$. Assume not, e.g. $\alpha \notin d(a)$. Then

$$-q_\alpha(a) = -q_{\alpha+1}(a) \geq q_{\alpha+1}(b) \geq b.$$

As $-q_\alpha(a) \in A_\alpha$, we get $-q_\alpha(a) \geq q_\alpha(b)$, by the definition of q_α. Consequently, $q_\alpha(a) \wedge q_\alpha(b) = 0$, contradiction. $\square$

As a first application, we can draw

Corollary 2.1.4 *If all A_α satisfy the ccc, then so does A.*

Indeed, assume by contradiction that there were an uncountable set C of pairwise disjoint elements in A. By 2.1.2 and the Δ-Lemma (cf. 0.0.2), we can assume that the family $\{d(c) : c \in C\}$ forms a Δ-system with kernel d say. Let δ be the biggest element of d and consider arbitrary $c_1, c_2 \in C$. By 2.1.3, from $c_1 \wedge c_2 = 0$ we get $q_{\alpha+1}(c_1) \wedge q_{\alpha+1}(c_2) = 0$ for some $\alpha \in d(c_1) \cap d(c_2) = d$. As $\delta \geq \alpha$ it follows that $q_{\delta+1}(c_1) \wedge q_{\delta+1}(c_2) = 0$. So $\{q_{\delta+1}(c) : c \in C\}$ consists of pairwise disjoint, non-zero (because $0 < c \leq q_{\delta+1}(c)$) elements of $A_{\delta+1}$, contradicting the ccc holding for that algebra. $\square$

In the second half of this section we prove the following principle.

Theorem 2.1.5 *Let $\mathcal{P}$ be a property of Boolean algebras that is preserved under well-ordered rc-filtrations, i.e. whenever $(A_\alpha)_{\alpha<\lambda}$ is a well-ordered rc-filtration of some Boolean algebra A and all A_α have the property $\mathcal{P}$, then A has property $\mathcal{P}$, too. Then $\mathcal{P}$ is preserved (in the same sense) under arbitrary σ-complete rc-filtrations.*

The proof needs the following result, sometimes called Iwamura's Lemma.

Lemma 2.1.6 *([29]) Every infinite directed set is the union of a continuous well-ordered chain of directed subsets of strictly smaller cardinality.*

Proof. Let J be the directed set in question. If J is countable, enumerate it: $\{j_n : n < \omega\}$. Put $k_0 = j_0$. Then choose by induction $k_{n+1} \in J$ in such a way that $j_n, k_n \leq k_{n+1}$. It is easy to see that $J_n = \{j_0, \dots j_n, k_0 \dots k_n, k_{n+1}\}$ yields the sought-for representation : $J = \bigcup_{n<\omega} J_n$.

Assume now that J is uncountable. Using that J is directed, we may define a binary operation $+$ on J such that $j_1 \leq j_1 + j_2$ and $j_2 \leq j_1 + j_2$. Consider the algebraic system $\langle J; + \rangle$ and represent it as the union of a continuous well-ordered chain of subalgebras of strictly smaller power. $\square$

Equipped with the lemma, we[1] now **prove** the theorem.

Let $(A_i)_{i \in I}$ be an rc-filtration of A such that all A_i have the property $\mathcal{P}$ and the partially ordered index set $(I, \leq)$ is σ-complete. We show that A has the property $\mathcal{P}$.

For a subset J of I we put $A_J = \bigcup_{i \in J} A_i$. In general, A_I is only a subset of A, but

CLAIM 1. *If J is directed, then $A_J \leq_{rc} A$.*

It is easy to see that A_J is a subalgebra of A. To prove relative completeness, we almost repeat the proof of 2.1.2 above. For $i \in I$ we let $q_i : A \rightarrow A_i$ denote the projections associated to $A_i \leq_{rc} A$. Notice that

(1) $i \leq j$ implies $q_i(a) \geq q_j(a)$, and

(2) if i is the supremum of the chain $(j_n)_{n < \omega}$, then $q_i(a) = q_{j_m}(a)$ for some $m < \omega$ (depending on a).

Let $a \in A$ be given and assume, by contradiction, that there is no least element in A_J above a. By induction we pick indices $j_n \in J$ such that

$$j_n \leq j_{n+1} \quad \text{and} \quad q_{j_n}(a) > q_{j_{n+1}}(a)$$

for all $n < \omega$. We can start with an arbitrary j_0. Having j_n, we know that $q_{j_n}(a)$ is not minimal in A_J above a, hence $q_{j_n}(a) > b \geq a$ for some $b \in A_J$. As $b \in A_j$ for some $j \in J$ and as J is directed, we find $j_{n+1} \geq j_n, j$ in J. Then $q_{j_n}(a) > b \geq q_{j_{n+1}}(a)$, as desired.

Now let i be the supremum (in I!) of the chain $(j_n)_{n < \omega}$. Then, by (1) and (2), $q_i(a) \leq q_{i_{m+1}}(a) < q_{i_m}(a) = q_i(a)$, for some m. This contradiction proves the claim.

The second claim will yield the theorem, for I itself is directed and $A = A_I$.

CLAIM 2. *If J is directed, then A_J has the property $\mathcal{P}$.*

This is proved by induction on the cardinality of J. If J is finite, then, being directed, it has a biggest element, j say. Clearly $A_J = A_j$, which has $\mathcal{P}$, by assumption.

If J is infinite, we use 2.1.6 to write it as $J = \bigcup_{\alpha < \kappa} J_\alpha$. By induction hypothesis, the algebras A_{J_α} have the property $\mathcal{P}$. They form a well-ordered filtration of A_J, which consists of relatively complete subalgebras, by claim 1. So A_J has the property $\mathcal{P}$, by assumption. $\square$

For later reference we single out the following immediate consequence of 2.1.4 and 2.1.5

Corollary 2.1.7 *A Boolean algebra having a σ-complete rc-filtration consisting of ccc subalgebras itself satisfies the ccc.*

---

[1] We have to thank S. Koppelberg for considerable simplifications of our original proof.

While the memory about Iwamura's Lemma is still fresh we draw another useful

Corollary 2.1.8 *Every skeleton is closed under unions of directed subfamilies.*

The **proof** is by induction on the cardinality of the directed subset and very similar to that of claim 2 above. The details can be safely left to the reader.

2.2   The Freese/Nation property

As mentioned in the introduction, the following property has been considered by R. Freese and J. B. Nation in connection with their investigations of projective lattices [14].

Definition 2.2.1 A Boolean algebra B is said to have the *Freese/Nation property*, abbreviated as (FN), if for each $b \in B$ there are two finite sets $U(b) \subseteq \{c \in B : b \leq c\}$ and $L(b) \subseteq \{c \in B : c \leq b\}$ such that, if $a \leq b$, then $U(a) \cap L(b) \neq \emptyset.\square$

There are two reformulations of (FN) which are often more convenient to work with. The letters I and S alude to 'interpolation' and 'separation', respectively. Below we choose the version most suited for the particular application calling all of them *the* Freese/Nation property.

Observation 2.2.2 *The property (FN) is equivalent to each of the following two.*

(FNI) *For each $b \in B$ there is a finite set $I(b) \subseteq B$ such that,*
  if $a \leq b$, then $a \leq x \leq b$ for some $x \in I(a) \cap I(b)$.

(FNS) *For each $b \in B$ there is a finite set $S(b) \subseteq B$ such that,*
  if $a \wedge b = 0$, then $a \leq c$ and $b \leq d$ for some disjoint $c, d \in S(a) \cap S(b)$.

If convenient, we treat U and L resp. I resp. S as mappings from B into the set of all finite subsets of B; $B \to [B]^{<\aleph_0}$ and say that they witness (FN) resp. (FNI) resp (FNS). To pass from one form to the other one uses the follwing formulas[2].

$$
\begin{array}{llll}
(FN) & \Rightarrow & (FNI): & I(b) = L(b) \cup U(b) \\
(FNI) & \Rightarrow & (FNS): & S(b) = I(b) \cup -I(-b) \\
(FNS) & \Rightarrow & (FN): & U(b) = S(b) \quad \text{and} \quad L(b) = -S(-b).
\end{array}
$$

The easy verifications are left to the reader.

. It is clear that all finite Boolean algebras have the Freese/Nation property, just take $I(b) = B$. To digest the definition, it may be helpful to look at the example of free Boolean algebras. Let F be freely generated by the set P. Then each element b of F belongs to some subalgebra $\langle P_b \rangle$, generated by a finite subset P_b of P and it is a matter of calculation to show that $I(b) = \langle P_b \rangle$ witnesses (FNI).

---

[2]Where, of course, $-M$ stands for $\{-m : m \in M\}$.

The reader has probably met these calculations in a course of propositional logic. Indeed, interpreting F as the algebra (modulo logical equivalence) of propositional formulas built up from the set P of propositional variables , we have that $a \leq b$ holds in F iff the formula $a \to b$ is a tautology. But then the *Interpolation Theorem* of propositional calculus says that there is a formula x containing only variables common to a and b (i.e. belonging to $P_a \cap P_b$) such that $a \to x$ and $x \to b$ are both tautologies.

Theorem 2.2.3 *The following assertions are equivalent for an arbitrary Boolean algebra A.*

(1) *A has the Freese/Nation property.*

(2) *A has an rc-skeleton.*

(3) $\{B \leq_{rc} A : |B| \leq \aleph_0\}$ *contains a club.*

(4) *A has a σ-complete rc-filtration consisting of countable subalgebras.*

In condition (4) the initiated reader will recognize the Boolean algebraic translation of Šĕpin's concept of open generation discussed in the introduction.

Proof of the theorem. The implication (1) $\Rightarrow$ (2) will easily follow from the

CLAIM *Let U and L witness (FN) for A. If $B \leq A$ has the property that*
$$U(b) \cup L(b) \subseteq B \text{ for all } b \in B, \text{ then } B \leq_{rc} A.$$

Indeed, let $a \in A$ be given. We shall find the greatest element of B below a. Denote $\bigvee \{c \in L(a) : c \in B\}$[3] by b. Clearly $b \in B$. As all elements of $L(a)$ are below a, we also have $b \leq a$.

To check that b is the sought-for greatest element of B below a, we consider $B \ni b' \leq a$. By (FN), there is some $c \in U(b') \cap L(a)$. The assumption on B yields $U(b') \subseteq B$, hence $c \in B \cap L(a)$ and c is an element of the join making up b. So $c \leq b$. On the other hand, being in $U(b')$, c must be above b'. Together the inequalities $b' \leq c$ and $c \leq b$ yield $b' \leq b$, as desired.

The claim is proved and it allows us to verify that (FN) implies the existence of an rc-skeleton: if U and L witness the property (FN) for A, then the fixed-point skeleton

$$S = \{B \leq A : U(b) \cup L(b) \subseteq B \text{ for all } b \in B\}$$

consists of relatively complete subalgebras.

The implication (2) $\Rightarrow$ (3) is straightforward, the countable elements of the skeleton form the desired club. (3) $\Rightarrow$ (4) is equally easy. Let $\mathcal{C}$ be the club in question. Considered under $\subseteq$, $\mathcal{C}$ is directed and σ-closed. So $(C)_{C \in \mathcal{C}}$ is the desired rc-filtration.

The proof of the implication (4) $\Rightarrow$ (1) will be based on the preparations of the previous section. The key to the proof is the following

---
[3] To be on the safe side: $\bigvee \emptyset = 0$.

Proposition 2.2.4 *A Boolean algebra has the property* (FNS) *if it has a well-ordered rc-filtration all members of which have the property* (FNS).

Proof. Denote the filtration by $(A_\alpha)_{\alpha < \tau}$ and the upper projections $A \to A_\alpha$ by q_α. Recall that the jump-set was defined as

$$d(a) = \{\alpha < \tau : q_\alpha(a) > q_{\alpha+1}(a)\}.$$

Let S_α witness (FNS) for A_α. We define S by setting

$$S(a) = S_0(q_0(a)) \cup \bigcup_{\alpha \in d(a)} S_{\alpha+1}(q_{\alpha+1}(a)).$$

By lemma 2.1.2, $S(a)$ is finite.

To check the crucial property, let us assume $a \wedge b = 0$. Then lemma 2.1.3 implies that either $q_0(a) \wedge q_0(b) = 0$ or $q_{\alpha+1}(a) \wedge q_{\alpha+1}(b) = 0$, for some $\alpha \in d(a) \cap d(b)$.

In the first case there are disjoint $c, d \in S(q_0(a)) \cap S(q_0(b))$ such that $q_0(a) \leq c$ and $q_0(b) \leq d$.

In the second case there are disjoint $c, d \in S(q_{\alpha+1}(a)) \cap S(q_{\alpha+1}(b))$ such that $q_{\alpha+1}(a) \leq c$ and $q_{\alpha+1}(b) \leq d$.

In both cases it follows that $a \leq c$ and $b \leq d$, as desired. $\square$

From this proposition we immediately get

CLAIM   *All countable Boolean algebras have the Freese/Nation property.*

Indeed, each countable Boolean algebra A has a filtration $(A_n)_{n < \omega}$ consisting of finite, hence relatively complete, subalgebras. Being finite, all A_n trivially have (FNS). By the proposition their union also has (FNS).

To end the proof of $(4) \Rightarrow (1)$ it remains to apply 2.1.5. The theorem is now completely proved. $\square$

Definition 2.2.5 A Boolean algebra will be called *rc-filtered* iff it satisfies one of the equivalent conditions of the above theorem. $\square$

We are not really happy with the expression 'rc-filtered'. In previous versions these algebras were called 'openly generated' and 'almost projective'. But both of these terms seem worse than the present one.

Immediately from 1.3.2(4) and the theorem above we get

Corollary 2.2.6 *Projective Boolean algebras are rc-filtered.*

It will take some time until we can prove that the two classes are actually distinct. Next we show that they coincide for Boolean algebras of cardinality $\leq \aleph_1$.

Corollary 2.2.7 *Assume that A is rc-filtered and let $M \subseteq A$ have power $\leq \aleph_1$. Then there exists a projective Boolean algebra B such that $M \subseteq B \leq_{rc} A$. It follows that A has the ccc and does not contain uncountable chains. Moreover, if A has cardinality $\leq \aleph_1$, then it is projective.*

Proof. Let S be an rc-skeleton of A. Considering the σ-complete rc-filtration $(S)_{S \in S_0}$, where S_0 is the collection of all countable elements of S (ordered under $\subseteq$), we may apply observation 1.1.6(2) to find a continuous subchain $(S_\alpha)_{\alpha < \omega_1}$ of S_0 such that $M \subseteq B = \bigcup_{\alpha < \omega_1} S_\alpha$. As the skeleton S is closed, $B \in S$, hence $B \leq_{rc} A$. As all S_α are countable, hence countably generated over each other, the chain witnesses the projectivity of B, by 1.3.2(6).

If M were an uncountable chain in A then B would contain an uncountable chain. But being projective, B is embeddable into a free Boolean algebra, and these are known to contain no uncountable chains. The ccc follows in the same way. $\square$

Notice that the ccc also follows from 2.1.7.

As all infinite complete Boolean algebras have uncountable chains, it follows from the corollary above that infinite rc-filtered algebras are not complete. We end this section with the first closedness results for the class of rc-filtered Boolean algebras, starting with a simple reformulation of 2.2.4.

Theorem 2.2.8 (Theorem 2.3 in [55])[4] *If a Boolean algebra has a well-ordered rc-filtration consisting of rc-filtered Boolean algebras, then it is itself rc-filtered.*

Proof. This is an immediate consequence of 2.2.4. $\square$

Together with a result to be proved in the next section the above theorem yields the following 'inductive' definition of the class.

Corollary 2.2.9 *All countable Boolean algebras are rc-filtered. An uncountable Boolean algebra is rc-filtered iff it has a well-ordered rc-filtration whose members are rc-filtered algebras of strictly smaller cardinality.*

Proof. 2.2.8 yields that Boolean algebras that have such a filtration are rc-filtered. Using an rc-skeleton it is not hard to find a well-ordered rc-filtration whose members have strictly smaller cardinality. In the next section we prove 2.3.1, which implies that its members are rc-filtered. $\square$

By 2.1.5, the previous theorem generalizes in the following way.

Corollary 2.2.10 (Theorem 2.5 in [55]) *If A has a σ-complete rc-filtration all elements of which are rc-filtered, then A is rc-filtered.*

The following result says that relative completeness can be relaxed to regularity or even σ-embeddings provided that the index set of the filtration is countable.

Theorem 2.2.11 (Bandlow, unpublished) *If A has a countable filtration $(A_i)_{i \in I}$ consisting of rc-filtered subalgebras such that $A_i \leq_\sigma A$ for all i, then A is rc-filtered.*

---

[4] As the Freese/Nation property is not available in the topological setting and Ščepin proved this result for general spaces, the topological counterpart is not as trivial as it looks here. In section 5.5 we reproduce Ščepins original argument.

It may be worth noticing that the condition $A_i \leq_\sigma A$ is satisfied if $A_i \leq_\sigma A_j$ for all $i \leq j$ and also if $A_i \leq_{reg} A$, resp. $A_i \leq_{reg} A_j$ for all $i \leq j$. The latter because of the Bockstein separation property (cf. next section), i.e. 1.2.20.

Proof. Choosing a cofinal subchain of type ω, the assertion reduces to the case that A is the union of a chain $(A_n)_{n<\omega}$ of rc-filtered subalgebras $A_n \leq_\sigma A$. For each $n < \omega$ and each $a \in A$ we fix a countable subset $F_n(a)$ of A_n which is closed under joins and generates $A_n \restriction a$. By assumption, there is an rc-skeleton, $\mathcal{S}_n$ say, for each A_n. Put

$$\mathcal{T}_n = \{T \leq A : T \cap A_n \in \mathcal{S}_n \text{ and } F_n(t) \subseteq T, \text{ for all } t \in T\}$$

and $\mathcal{T} = \bigcap_{n<\omega} \mathcal{T}_n$. By 1.1.4 and 1.1.7, each $\mathcal{T}_n$ is a skeleton. So $\mathcal{T}$ is a skeleton, by 1.1.3. To see that all $T \in \mathcal{T}$ are relatively complete, consider any $a \in A$. Then $a \in A_n$ for some n. As $T \cap A_n \in \mathcal{S}_n$, we have $T \cap A_n \leq_{rc} A$. This allows us to pick an element t_n which is minimal in $T \cap A_n$ above a. To show that this t_n is, in fact, minimal in the whole of T, consider any $s \in T$ such that $s \geq a$. Then $a \in A_n \restriction s$, hence $a \leq s' \leq s$, for some $s' \in F_n(s) \subseteq T \cap A_n$. As t_n was minimal in $T \cap A_n$ above a, we must have $t_n \leq s'$, hence $t_n \leq s$, as desired. $\square$

Topological duality: The dual spaces of rc-filtered Boolean algebras are what Ščepin called 'openly generated'. In section 2.9 below we explain, why these spaces are also called κ-metrizable.

2.3 Subalgebras and the Bockstein separation property

The main result of this section will be a necessary and sufficient condition for a subalgebra of an rc-filtered Boolean algebra to be rc-filtered again. But we begin with an easy to prove but useful sufficient condition.

Theorem 2.3.1 *If $A \leq_{rc} B$ and B is rc-filtered, then A is also rc-filtered.*

In the **proof** we use the Freese/Nation property. Let I_B witness (FNI) for B and denote the projection $B \to A$ by q, as usual. For $a \in A$ we define $I_A(a) = q(I_B(a))$. and verify that I_A witnesses (FNI) for A. Finiteness is clear. If $a_1 \leq a_2$, then $a_1 \leq x \leq a_2$ for some $x \in I_B(a_1) \cap I_B(a_2)$. Obviously, $a_1 \leq x \leq q(x)$. And from $x \leq a_2 \in A$, we get $q(x) \leq a_2$, by minimality of $q(x)$ above x. Taken together the two inequalities yield what we need, $a_1 \leq q(x) \leq a_2$, with

$$q(x) \in q(I_B(a_1) \cap I_B(a_2)) \subseteq q(I_B(a_1)) \cap q(I_B(a_2)) = I_A(a_1) \cap I_A(a_2). \quad \square$$

It is known that relatively complete subalgebras of projective Boolean algebras need not be projective. That shows that the two classes are different. Examples are not easy to obtain; we present one in section 3.4.

In order to formulate the necessary and sufficent condition we need the following

Definition 2.3.2 A Boolean algebra A is said to have the *Bockstein separation property*, abbreviated as (BSP), if all regular ideals of A are countably generated.

The dual topological property is that all regular open sets are of type F_σ. The name has been introduced by Pełczynski motivated by Bockstein's result that arbitrary products of separable metrizable spaces have this property. Šcepin reinvented the concept and realized its importance for the theory under discussion. He called it perfect κ-normality. Here is why we speak of a separation property.

Exercise. Prove that A has the (BSP) if for each pair I, J of ideals of A such that $I \cap J = \{0\}$ there is a pair I_1, J_1 of countably generated ideals such that $I \subseteq I_1$, $J \subseteq J_1$ and $I_1 \cap J_1 = \{0\}$.

Proposition 2.3.3 *All rc-filtered Boolean algebras have the* (BSP).

Proof. Let $(A_i)_{i \in I}$ be a σ-complete rc-filtration of A with all A_i countable. Let J be a regular ideal of A. Then, because of the ccc, we can write $J = M^d$ for some countable set M (cf. 1.2.15). By 1.1.6, there is some index $i \in I$ such that $M \subseteq A_i$. It will be sufficient to verify that J is generated by its intersection with this A_i. To do so, let us consider an arbitrary $a \in J$. Let $\bar{a}$ be minimal in A_i above a. It is sufficient to establish $\bar{a} \in J = M^d$. But, if that were not the case, then $\bar{a} \wedge m > 0$ for some $m \in M$. As $a \in J = M^d$, we would have $a \leq \bar{a} - m < \bar{a}$, contradicting the minimality of $\bar{a}$. $\square$

Next we produce an example, which is originally due to Engelking (cf. also 1.14 in [35]). It shows that we cannot replace relative completeness by regularity in the theorem above. Let A be an atomless Boolean algebra and let U and V be distinct ultrafilters of A.

CLAIM 1. $B = \{a \in A : a \in U \Leftrightarrow a \in V\}$ *is a dense subalgebra of* A.

We leave it as an exercise to check that B is, indeed, a subalgebra (in fact $B = (U \cap V)$) of A. To prove density, take any non-zero $a \in A$. If $a \notin B$, then $a \in U$ and $a \notin V$, say. Take two non-zero disjoint a_1, a_2 below a. At most one of these can belong to U. Not belonging to V either, the other is in B.

Consider any $a \in U \setminus V$ so $a \notin B$. As B is dense hence regular in A, the ideal $B \restriction a$ is regular. Let $X \subseteq B \restriction a$ generate this ideal and assume that X is closed under finite unions (which does not increase the cardinality of X).

CLAIM 2. $\{a - x : x \in X\}$ *generates the ultrafilter* U.

Indeed, all elements $a - x$ belong to U. For, $x \leq a \notin V$ implies $x \notin V$. Therefore, from $x \in B$, we get $x \notin U$. Hence $-x \in U$ and $a - x = a \wedge -x \in U$.

Consider now an arbitrary $u \in U$. We find some $x \in X$ such that $a - x \leq u$ or, equivalently, $a - u \leq x$. From $a - u \notin U$ and $a - u \leq a \notin V$, we get that

$a - u$ does not belong to V either. Hence $a - u \in B\!\restriction a$. Consequently, $a - u \leq x$ for some $x \in X$, as desired.

Now we become more specific.

Example 2.3.4 *Let us take* $\mathrm{Fr}\,\omega_1$ *for* A *and arbitrary* U *and* V. *Then* B *is a dense subalgebra of* $\mathrm{Fr}\,\omega_1$, *which is not rc-filtered, not having the* (BSP).

Indeed, claim 2 shows that B has many regular ideals that are not countably generated, for, all ultrafilters of $\mathrm{Fr}\,\omega_1$ have character $\aleph_1$ (this is well-known, cf. also 2.7.5). Notice that all ultrafilters of B also have character $|B| = \aleph_1$ (exercise). □

Finally we show that the (BSP) really is what matters.

Theorem 2.3.5 *Assume that* $A \leq B$ *and that* B *is rc-filtered. Then* A *is rc-filtered if and only if it has the* (BSP).

Proof. One direction trivially follows from the proposition above. So let us assume that A has the (BSP). Use Zorn's Lemma to choose an ideal $J \subseteq B$ maximal with respect to the property $J \cap A = \{0\}$. Imediately from 1.2.21 we get

CLAIM 1.   *For each* $b \in B$, *the set*

$$A\!\restriction (b \vee J) = \{a \in A : a \leq b \vee j \text{ for some } j \in J\} = \{a \in A : a - b \in J\}$$

   is a regular ideal of A.

The (BSP) then allows us to take a mapping $F : B \to [A]^{\leq \aleph_0}$ such that $F(b)$ generates $A\!\restriction (b \vee J)$.

CLAIM 2.   *If* $S \leq_{rc} B$ *and* S *is closed under* F, *then* $S \cap A \leq_{rc} A$.

Indeed, let $a \in A$ be given and take $s \in S$ minimal above a. Then $a \in A\!\restriction (s \vee J)$. Using that $F(s) \subseteq S \cap A$ we find elements $a' \in S \cap A$ and $j \in J$ such that $a \leq a' \leq s \vee j$.

To see that a' is the sought-for minimal element of $S \cap A$ above a, we consider any $t \in S \cap A$ such that $a \leq t$. Then $s \leq t$, by minimality of s (and $t \in S$), hence $a' \leq s \vee j \leq t \vee j$. Consequently, $a' - t \leq j \in J$. As $a' - t$ is also in A, and $A \cap J = \{0\}$, we get $a' - t = 0$, i.e. $a' \leq t$, as desired. The claim is proved.

Now we prove the theorem from claim 2. By assumption, there is an rc-skeleton, $\mathcal{S}$ say, for B. Intersecting if necessary with the fixed-point skeleton of F, we can assume that all $S \in \mathcal{S}$ are closed under F. Then $(A \cap S)_{S \in \mathcal{S}}$ is a σ-complete filtration of A, which, by claim 2, consists of relatively complete subalgebras. It shows that A is rc-filtered. □

Corollary 2.3.6 *If* $A \leq_\sigma B$ *and* B *is rc-filtered, then* A *is also rc-filtered.*

Indeed, this is an immediate consequence of the theorem and the following

Observation 2.3.7 *If $A \leq_\sigma B$ and B has the (BSP), then A also has the (BSP).*

To see this, let I be a regular ideal of A. Then $I = M^{d_A}$ for some $M \subseteq A$. Then $J = M^{d_B}$ is a regular ideal of B, hence countably generated, by the (BSP). From $A \leq_\sigma B$ it now follows that $I = J \cap A$ is also countably generated. $\square$

2.4 Homomorphic images

We know already that not all Boolean algebras are rc-filtered. As all Boolean algebras are homomorphic images of free Boolean algebras, not all homomorphisms preserve the property of being rc-filtered. But some do, as we now see. The following proposition is a special case of theorem 6 in [53], where Ščepin deals with adequate pairs in general.

Proposition 2.4.1 *Assume that A is rc-filtered. If the ideal I is regular or countably generated, then the quotient A/I is also rc-filtered.*

Proof. By 2.3.3, regular ideals are countably generated, so there is only one case to consider. Let $I_0 \subseteq I$ be countable and generate I. We argue by induction on $|A|$. If A is countable, then A/I is also countable and rc-filtered anyway. If A is uncountable, we use 2.2.9 to find an rc-filtration $(A_\alpha)_{\alpha < \kappa}$ consisting of rc-filtered subalgebras of power strictly less than $|A|$. Moreover, we can assume that $I_0 \subseteq A_0$. It is clear that $(A_\alpha/I)_{\alpha < \kappa}$ is a filtration of A/I, which, by induction, consists of rc-filtered Boolean algebras. By 2.2.8 it will be sufficient to check that $A_\alpha/I \leq_{rc} A/I$, for all α. Let $a/I \in A/I$ be given. As $A_\alpha \leq_{rc} A$, we may take the maximal element b, say of A_α below a. Let us check that b/I is maximal below a/I. Suppose $c/I \in A_\alpha/I$ is below a/I. Then $c - i \leq a$ for some $i \in I$. As $I_0 \subseteq A_\alpha$ generates I, we can assume $i \in A_\alpha$. But then $c - i \in A_\alpha$, hence $c - i \leq b$ and, therefore, $c/I \leq b/I$. $\square$

The reader will have no difficulty to adapt the above proof to yield the same result for projective Boolean algebras (cf. 1.7 in [35]).

Now we want to characterize those homomorphisms that preserve the class of rc-filtered Boolean algebras.

For comparison we first reformulate two results about projective Boolean algebras in somewhat unusual form. Let us call a homomorphism $\varphi : A \to B$ *invertible* if it has a right inverse, which is a homomorphism, i.e. if there is a homomorphism $\varepsilon : B \to A$ such that $\varphi \circ \varepsilon = id_B$.

Observations 2.4.2 *Let $\varphi : A \to B$ be a surjective homomorphism.*

(1) *If A is projective and φ is invertible, then B is projective.*

(2) *If B is projective, then φ is invertible.*

Indeed, the second assertion is an immediate consequence of the definition of projectivity. The first one follows from 1.3.2(2) and the fact that being a retract is a transitive property.

Širokov's characterization 1.3.2(3) implies that we can strengthen the first of these assertions by demanding less of the inverse of φ.

Proposition 2.4.3 ([56]) *Let $\varphi : A \to B$ be a surjective homomorphism. If A is projective and there is a mapping $\varepsilon : B \to A$ preserving 0 and $\wedge$ such that $\varphi \circ \varepsilon = id_B$, then B is projective.*

To formulate similar results for rc-filtered Boolean algebras, we introduce the following shorthand.

Definition 2.4.4 A homomorphism $\varphi : A \to B$ will be called $\perp$-*invertible* ($\leq$-*invertible*) if there is a mapping $\varepsilon : B \to A$ such that $\varphi \circ \varepsilon = id_B$ and ε preserves disjointness (is monotone), i.e. $b_1 \wedge b_2 = 0$ implies $\varepsilon(b_1) \wedge \varepsilon(b_2) = 0$ ($b_1 \leq b_2$ implies $\varepsilon(b_1) \leq \varepsilon(b_2)$), for all $b_1, b_2 \in B$. $\square$

The next result is due to Širokov and can be found (in topological form) in [56]. Our proof is much easier than the original one, due to the Freese/Nation property.

Theorem 2.4.5 *Let $\varphi : A \to B$ be a surjective homomorphism.*

(1) *If A is rc-filtered and φ is $\perp$-invertible, then B is rc-filtered.*

(2) *If B is rc-filtered, then φ is $\perp$-invertible.*

Proof of (1). Let $\varepsilon : B \to A$ be the $\perp$-inverse of φ. Let S_A witness (FNS) for A. We define $S_B(b) = \varphi(S_A(\varepsilon(b))$ and check that this mapping witnesses (FNS) for B. Finiteness is clear. Suppose that $b_1 \wedge b_2 = 0$. Then $\varepsilon(b_1) \wedge \varepsilon(b_2) = 0$. It follows that there are disjoint $c, d \in S_A(\varepsilon(b_1)) \cap S_A(\varepsilon(b_2))$ such that $\varepsilon(b_1) \leq c$ and $\varepsilon(b_2) \leq d$. As φ is a homomorphism, $\varphi(c), \varphi(d) \in S_B(b_1) \cap S_B(b_2)$ are disjoint and

$$b_1 = \varphi(\varepsilon(b_1)) \leq \varphi(c), \quad \text{and} \quad b_2 = \varphi(\varepsilon(b_2)) \leq \varphi(d),$$

as desired.

The proof of (2) starts with a

CLAIM　　*There is a right inverse $\delta : B \to A$ of φ such that $\delta(-b) = -\delta(b)$ for all $b \in B$.*

This has nothing to do with rc-filtrations. Indeed, as φ is surjective, there exists some right inverse $\theta : B \to A$ of φ. Let U be any ultrafilter of B and put

$$\delta(b) = \begin{cases} \theta(b), & \text{if } b \in U \\ -\theta(-b), & \text{if } b \notin U. \end{cases}$$

This δ is easily checked to be as desired.

Now let $S : B \rightarrow [B]^{<\aleph_0}$ witness (FNS) for B and put

$$\varepsilon(b) = \delta(b) \wedge \bigwedge \{\delta(c) : b \leq c \text{ and } c \in S(b) \text{ or } -c \in S(b)\}.$$

Then

$$\varphi(\varepsilon(b)) = \varphi(\delta(b)) \wedge \bigwedge \{\varphi(\delta(c)) : \ldots\} = b \wedge \bigwedge \{c : \ldots\} = b,$$

because all $c \in \{\ldots\}$ are above b. Moreover, $b_1 \wedge b_2 = 0$ implies the existence of some disjoint $c, d \in S(b_1) \cap S(b_2)$ such that $b_1 \leq c$ and $b_2 \leq d$. From $b_2 \leq d \leq -c$ and $--c = c \in S(b_2)$ it follows that $\delta(-c)$ is a meetand in the definition of $\varepsilon(b_2)$, hence $\varepsilon(b_1) \wedge \varepsilon(b_2) \leq \delta(c) \wedge \delta(-c) = \delta(c) \wedge -\delta(c) = 0$, by the choice of δ.
□

Now it is a matter of some diagram manipulation to get the following characterization of our class, which parallels the definition of projectivity.

Corollary 2.4.6 *A Boolean algebra A is rc-filtered iff for all homomorphisms $A \xrightarrow{\varphi} B \xleftarrow{\psi} C$, with ψ surjective, there exists a $\bot$-preserving mapping $\varepsilon : A \rightarrow C$ such that $\psi \circ \varepsilon = \varphi$.*

The diagram is the same as with projectivity (cf. page 22), only that ε need not be a homomorphism any more, but preserves only part of the structure. (Compare also with 1.3.2(3).)

Proof.   Let ψ be a surjective homomorphism of a free Boolean algebra F onto A. Applying the diagram property to $A \xrightarrow{id} A \xleftarrow{\psi} F$ shows that ψ is $\bot$-invertible. So, A is rc-filtered, by 2.4.5(1).

For the other direction we assume that A is rc-filtered and consider homomorphisms $A \xrightarrow{\varphi} B \xleftarrow{\psi} C$. [5]

Let $\theta : F \rightarrow A$ be a homomorphism of the free Boolean algebra F onto A. By 2.4.5(2), there exists a $\bot$-inverse $\beta : A \rightarrow F$, say.

Using the surjectivity of ψ, we can pick an element $c_x \in C$ for each free generator x of F such that $\varphi(\theta(x)) = \psi(c_x)$. Let $\delta : F \rightarrow C$ denote the homomorphism extending the mapping $x \mapsto c_x$. Then $\psi \circ \delta = \varphi \circ \theta$, because this is true for all generators of F.

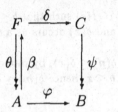

---

[5] An alternative to the following diagram manipulations is to choose any mapping $\theta : A \rightarrow C$ such that $\psi \circ \theta = \varphi$, which exists by the surjectivity of ψ. Then δ and ε can be defined as in the proof of 2.4.5(2).

The mapping $\varepsilon = \delta \circ \beta$ will be as desired. Indeed, as δ is a homomorphism, ε preserves as much of the structure as β does. Moreover, from $\psi \circ \delta = \varphi \circ \theta$ and $\theta \circ \beta = id_A$ we get

$$\psi \circ \varepsilon = \psi \circ \delta \circ \beta = \varphi \circ \theta \circ \beta = \varphi \circ id_A = \varphi$$

as desired. $\square$

An analogon of Sirokov's theorem holds for monotone mappings. It is essentially due to Freese and Nation (Theorem 2 of [14], which deals with lattices instead of Boolean algebras).

Theorem 2.4.7 *Let $\varphi : A \to B$ be a surjective homomorphism.*

(1) *If A is rc-filtered and φ is $\leq$-invertible, then B is rc-filtered.*

(2) *If B is rc-filtered, then φ is $\leq$-invertible.*

Proof.   (1) Let $\varepsilon : B \to A$ be the monotone inverse of φ and let I_A witness (FNI) for A. A routine verification then shows that I_B defined by $I_B(b) = \varphi(U_A(\varepsilon(b))$ witnesses (FNI) for B.

(2) Now let I witness (FNI) for B. We have to construct a $\leq$-inverse of φ. Let us call a monotone mapping $\delta : M \to A$ *suitable* if $M \subseteq B$ and $\varphi(\delta(m)) = m$ for all $m \in M$. Then what we need is a suitable mapping defined on the whole of B. The main step in the construction is singled out in the following

CLAIM 1. *If M is closed under I and N is finite, then each suitable $\delta :$ $M \cup N \to A$ can be extended to a suitable $\bar{\delta} : M \cup N \cup \{b\} \to A$, for whatever $b \in B$.*

Indeed, there is nothing to show, if b happens to be in $M \cup N$. Otherwise, we use the surjectivity of φ to choose some $a \in A$ such that $\varphi(a) = b$. Then we put

$$\bar{\delta}(b) = \quad [a \ \wedge \ \bigwedge\{\delta(n) : b \leq n \in N\} \ \wedge \ \bigwedge\{\delta(m) : b \leq m \in M \cap I(b)\}]$$
$$\vee \ [\bigvee\{\delta(d) : b \geq n \in N\} \ \vee \ \bigvee\{\delta(m) : b \geq m \in M \cap I(b)\}].$$

Then $\varphi(\bar{\delta}(b)) = b$ is easily verified. To check that $\bar{\delta}$ remains monotone, one has to distinguish several cases. We just consider the most complicated of them. Namely, given $b \leq e \in M$ we show $\bar{\delta}(b) \leq \bar{\delta}(e) = \delta(e)$.

As I witnesses (FNI), there is some $x \in I(b) \cap I(e)$ such that $b \leq x \leq e$. By assumption, $I(e) \subseteq M$, hence $b \leq x \in I(b) \cap M$ and $\delta(x)$ occurs as a meetand in the definition of $\bar{\delta}(b)$.

Moreover, if $b \geq n \in N$, then $n \leq x$, hence $\delta(n) \leq \delta(x)$, by monotonicity of δ. Analogously, $b \geq m \in I(b) \cap M$ implies $m \leq b \leq x$, hence $\delta(m) \leq \delta(x)$.

Putting together these estimates we get

$$\bar{\delta}(b) \ = [a \wedge \ldots \delta(x) \ldots] \vee \bigvee\{\ldots \delta(m) \ldots\} \vee \bigvee\{\ldots \delta(n) \ldots\}$$
$$\leq \delta(x) \vee \delta(x) \vee \delta(x) \leq \delta(e),$$

by monotonicity of δ again. The claim is proved. Using it to extend δ from M to $M \cup \{n_0\}$ to $M \cup \{n_0, n_1\}$ to $M \cup \{n_0, n_1, n_2\}$ to ... yields

CLAIM 2. *Let M be closed under I. Then each suitable $\delta : M \to A$ has a suitable extension $\bar{\delta} : M \cup N \to A$ for each countable $N \subseteq B$.*

To finish the proof we consider the collection $\mathcal{P}$ of all suitable mappings defined on subsets of B that are closed under I. We order $\mathcal{P}$ by inclusion

$$\gamma \leq \delta \Leftrightarrow \text{dom}\gamma \subseteq \text{dom}\delta \text{ and } \delta\lceil \text{dom}\gamma = \gamma.$$

By Zorn's Lemma, there is a maximal element, δ say, in $\mathcal{P}$. Denote its domain by M. We are done if $M = B$. So assume $b \notin M$. Choose a countable set N which is closed under I and contains b. Then, by claim 2, δ can be extended to an element of $\mathcal{P}$, which is defined on $M \cup N$. This contradicts the maximality of δ and ends the proof. $\square$

As in the $\perp$-case we can get a diagram property of the theorem just proved.

Corollary 2.4.8 *A Boolean algebra A is rc-filtered iff for all homomorphisms $A \xrightarrow{\varphi} B \xleftarrow{\psi} C$, with ψ surjective, there exists a monotone mapping $\varepsilon : A \to C$ such that $\psi \circ \varepsilon = \varphi$.*

Assuming that $\varphi : A \to B$ is a homomorphism onto an rc-filtered Boolean algebra we have learned to construct a $\perp$-inverse and a $\leq$-inverse. In general, these two mappings will be distinct. In section 3.2 we will be able to prove that, in fact, there is a right inverse of φ wich is monotone *and* $\perp$-preserving at the same time.

2.5 Products

For completeness sake we include this short section on products. All results are very easy to prove; we often confine ourselves to giving hints.

Proposition 2.5.1 *The product $\prod_{i \in I} A_i$ is rc-filtered iff I is finite and all factors A_i are rc-filtered.*

Proof. The index set must be finite, for, otherwise, the power-set algebra on I would embed into $\prod_{i \in I} A_i$ which would, therefore, contain uncountable chains. All A_i have to be rc-filtered being quotients of $\prod_{i \in I} A_i$ by principal, hence countably generated, ideals.

For the other direction it is sufficient and easy to prove that

$$S \times T = \{S \times T : S \in \mathcal{S}, T \in \mathcal{T}\}$$

is an rc-skeleton for $A \times B$ provided that $\mathcal{S}$ and $\mathcal{T}$ are rc-skeletons for A and B, respectively. $\square$

Recall that

Definition 2.5.2 *The weak product $\prod_{i \in I}^{w} A_i$ of a family $(A_i)_{i \in I}$ is the subalgebra of the full product $\prod_{i \in I} A_i$ which consists of all $(a_i)_{i \in I}$ for which either $\{i : a_i \neq 0\}$ or $\{i : a_i \neq 1\}$ is finite. $\square$*

Proposition 2.5.3 *The weak product $\prod_{i \in I}^{w} A_i$ is rc-filtered iff I is countable and all A_i are rc-filtered.*

Proof. The index set I must be countable because of the ccc. As quotients by principal ideals, all A_i must be rc-filtered.

To prove the other direction we assume $I = \omega$ and fix rc-skeletons $\mathcal{S}_n$ for all A_n. It will be sufficient to check that

$$\prod_{n < \omega}^{w} \mathcal{S}_n = \{\prod_{n < \omega}^{w} S_n : S_n \in \mathcal{S}_n\}$$

is an rc-skeleton for $\prod_{n < \omega}^{w} A_n$.

CLOSEDNESS. Let $(\prod_{n < \omega}^{w} S_n^{\alpha})_{\alpha < \kappa}$ be a chain. Then all $(S_n^{\alpha})_{\alpha < \kappa}$ are chains and it is sufficient to check

$$\bigcup_{\alpha < \kappa} (\prod_{n < \omega}^{w} S_n^{\alpha}) = \prod_{n < \omega}^{w} (\bigcup_{\alpha < \kappa} S_n^{\alpha}).$$

The inclusion from left to right is obvious. For the opposite inclusion we take an arbitrary $s = (s_n) \in \prod_{n < \omega}^{w} (\bigcup_{\alpha < \kappa} S_n^{\alpha})$. For each $n < \omega$ we take the least $\alpha(n)$ such that $s_n \in S_n^{\alpha}$. As almost all s_n belong to $\{0,1\} \subseteq S_n^{0}$, there is some $\beta < \kappa$ bigger than all $\alpha(n)$. Clearly, $s \in \prod_{n < \omega}^{w} S_n^{\beta}$.

ABSORPTION. Consider any $C \leq \prod_{n < \omega}^{w} A_n$. Let C_n denote the projection of C onto A_n. Then $C_n \leq A_n$ and we can take $S_n \in \mathcal{S}_n$ such that $C_n \leq S_n$ and $|S_n| \leq |C_n| \leq |C|$. It follows that $C \leq \prod_{n < \omega}^{w} S_n$ and $|\prod_{n < \omega}^{w} S_n| \leq \sup |S_n| \leq |C|$. (Here we use the countability assumption!)

RELATIVE COMPLETENESS. To show that $S_n \leq_{rc} A_n$ implies $\prod_{n < \omega}^{w} S_n \leq \prod_{n < \omega}^{w} A_n$, we consider any $a = (a_n)_{n < \omega} \in \prod_{n < \omega}^{w} A_n$. For each n, we let s_n be maximal in S_n below a_n. As $a_n \in \{0,1\}$ for almost all n, we have $s_n = a_n$ for almost all n. It follows that s belongs to $\prod_{n < \omega}^{w} S_n$. It is obvious that s is maximal below a. $\square$

We now turn to free products. From 1.2.6 it follows easily that

$$S \otimes T = \{S \otimes T : S \in \mathcal{S}, T \in \mathcal{T}\}$$

is an rc-skeleton for $A \otimes B$ provided that $\mathcal{S}$ and $\mathcal{T}$ are rc-skeletons of A and B, respectively.

Proposition 2.5.4 (Theorem 15 in [53]) *The class of rc-filtered Boolean algebras is closed under arbitrary free products.*

Proof. Let $(A_\alpha)_{\alpha < \kappa}$ consist of rc-filtered Boolean algebras. Then for all limit ordinals $\gamma < \kappa$

$$\bigotimes_{\alpha < \gamma} A_\alpha = \bigcup_{\beta < \gamma} \bigotimes_{\alpha < \beta} A_\alpha.$$

Using that this filtration is rc, one proves by induction on γ that all $\bigotimes_{\alpha < \gamma} A_\alpha$ are rc-filtered. The above observation settles successor steps and 2.2.8 limit steps. $\square$

As an **exercise** the reader may prove that all results of this section hold for projective Boolean algebras as well (cf. section 1 of [35]).

2.6 First results about cardinal functions

Our standard reference for cardinal functions on Boolean algebras is [42]. All unexplained notions and notations can be found there; most of them also in [34]. Let us just recall some definitions and introduce some notations that differ slightly from [42].

To avoid possible trivialities, we tacitly assume all Boolean algebras occuring in this and the next section to be infinite and all filters and ideals to be proper.

Let A be a Boolean algebra. If F is a filter of A, then

$$\chi(F, A) = \min\{|H| : H \subseteq F \text{ generates } F\}$$

is called the *character* of F in A. The cardinal number

$$\chi(A) = \sup\{\chi(U, A) : U \in Ult\, A\}$$

is called the *character* of A. Sometimes it will be technically easier to work with ideals rather than with filters. Therefore, we define

$$\gamma(I, A) = \min\{|H| : H \subseteq I \text{ generates } I\}$$

for every ideal of $I \subseteq A$. Both functions are connected in the obvious way: if F is a filter of A, then $-F = \{-f : f \in F\}$ is an ideal and

$$\chi(F, A) = \gamma(-F, A).$$

The same procedure works the other way round, i.e.

$$\gamma(I, A) = \chi(-I, A).$$

In particular, $\chi(A) = \sup\{\gamma(P, A) : P \subseteq A \text{ is a prime ideal}\}$.

Let M and N be subsets of A. N is called *dense* in M if for all non-zero $m \in M$ there is a non-zero $n \in N$ such that $n \leq m$. Notice that we do not require $N \subseteq M$. For $M \subseteq A$ we put

$$\pi(M, A) = \min\{|N| : N \subseteq A \text{ is dense in } M\}.$$

The cardinal $\pi(A, A)$ is usually written $\pi(A)$ and called the π-*weight* of A. The π-*character* of A is the number

$$\pi\chi(A) = \sup\{\pi(U, A) : U \in Ult\, A\}.$$

Recall that the cardinal number κ is called a *caliber* of A if each $X \subseteq A$ of power κ contains a subset Y still of power κ which has the *finite intersection property*, i.e. $\bigwedge Y_0 \neq 0$ for all finite $Y_0 \subseteq Y$.

All results on cardinal functions are taken from [54]. They generalize corresponding results for projective Boolean algebras (part of which can be found in [35]). These, in turn, are often special cases of results concerning subalgebras of free Boolean algebras.

All proofs of this section are based on the same idea. They only use that the family of relatively complete subalgebras is absorbing.

Proposition 2.6.1 *If A is rc-filtered, then $\pi(A) = |A|$.*

Proof. Take a dense subset N of A such that $|N| = \pi(A)$. Using that A is rc-filtered, we find some $D \leq_{rc} A$ such that $N \subseteq D$ and $|D| \leq |N|$.

We check that $A = D$. To see this take any $a \in A$ and a maximal $d \in D$ below a. If $a - d$ were non-zero, then $0 < n < a - d$ for some $n \in N \subseteq D$, contradicting the maximality of d. It follows that $a = d \in D$. We have proved that $|A| = |D| \leq \pi(A)$. The reverse is obvious. $\square$

Proposition 2.6.2 (Theorem 8 in [54]) *If A is rc-filtered, then $\pi(U, A) = \chi(U, A)$ for all $U \in Ult\, A$.*

Proof. $\pi \leq \chi$ is obvious. To prove the other direction, we consider any $N \subseteq A$ that is dense in U and has power $\pi(U, A)$. There is a subalgebra B of A which has power $|N|$ such that $N \subseteq B \leq_{rc} A$.

We show that $B \cap U$ generates U. Take $a \in U$ and consider $b \in B$ maximal below a. It will be sufficient to establish $b \in U$. If that were not true, then $a - b \in U$, hence $0 < n \leq a - b$ for some $n \in N \subseteq B$, contradicting the maximality of b. It follows that $\chi(U, A) \leq |B \cap U| \leq |B| = |N| = \pi(U, A)$. $\square$

In the rest of this section we prove some assertions that will be useful in the next section.

Lemma 2.6.3 *Assume that $A \leq B$, where B is rc-filtered. Then each ultrafilter U of A extends to a proper filter F of B such that $\chi(F, B) \leq \pi(U, A)$.*

Proof. Take $N \subseteq A$ dense in U such that $|N| = \pi(U, A)$. As B is rc-filtered, there is a subalgebra D of B such that $N \subseteq D \leq_{rc} B$ and $|D| = |N|$. Let $p : B \to D$ denote the lower projection and consider $G = \{p(a) : a \in U\}$. From $p(a) \leq a$ it follows that the filter generated by G in B contains U. As $|G| \leq |D| = \pi(U, A)$, it will be sufficient to prove that G generates a *proper* filter of B, which is equivalent to G having the finite intersection property.

Consider any $a_1, \ldots, a_n \in U$. Then $a_1 \wedge \ldots \wedge a_n \in U$, hence $0 < d \leq a_1 \wedge \ldots \wedge a_n$ for some $d \in N \subseteq D$. It follows that $d \leq a_i$ for all i, hence $d \leq p(a_i)$, by the definition of p. So, $0 < d \leq p(a_1) \wedge \ldots \wedge p(a_n) \neq 0$, as was to be shown. $\square$

For the readers convenience we recall the concept of jump set from section 2.1. It refers to an rc-filtration $(A_\alpha)_{\alpha < \tau}$ of a Boolean algebra A with $q_\alpha : A \to A_\alpha$ denoting the upper projections. The jump set of $a \in A$ was defined as

$$d(a) = \{\alpha < \tau : q_\alpha(a) > q_{\alpha+1}(a)\}$$

and shown to be finite for all $a \in A$. Here we need the following generalization of 2.1.3.

Lemma 2.6.4 (Lemma 5.3 in [54]) *Let $X \subseteq A$ and $\beta < \tau$ be such that[6]*

$$d(x) \cap d(y) \cap [\beta, \tau) = \emptyset$$

---

[6] As usual, $[\beta, \tau)$ denotes the interval $\{\alpha : \beta \leq \alpha < \tau\}$.

for all $x \neq y \in X$. If $\{q_\beta(x) : x \in X\}$ has the finite intersection property, then X also has the finite intersection property.

Proof. Let $x_1, \ldots, x_n \in X$ be given. We show by induction on $\alpha < \tau$ that

$$q_\alpha(x_1) \wedge \ldots \wedge q_\alpha(x_n) \neq 0.$$

For $\alpha \leq \beta$ this follows from $q_\alpha \geq q_\beta$ and the assumption.

Let α be a LIMIT ordinal. Then, for each $i \leq n$ there is $\delta_i < \alpha$ such that $q_\alpha(x_i) = q_\delta(x_i)$ for all $\delta \in (\delta_i, \alpha)$. Taking $\delta < \alpha$ bigger than all δ_i, we get

$$q_\alpha(x_1) \wedge \ldots \wedge q_\alpha(x_n) = q_\delta(x_1) \wedge \ldots \wedge q_\delta(x_n).$$

The right-hand side is non-zero, by induction hypothesis.

Let us now consider the SUCCESSOR STEP. The assumption on the jump sets implies that at most one x_i, x_1 say, jumps at α. Consequently,

$$a = q_{\alpha+1}(x_2) \wedge \ldots \wedge q_{\alpha+1}(x_n) = q_\alpha(x_2) \wedge \ldots \wedge q_\alpha(x_n)$$

belongs to A_α. By induction hypothesis, $a \wedge q_\alpha(x_1) \neq 0$. As $q_\alpha(x_1)$ is minimal in A_α above x_1, it follows that $a \wedge x_1 \neq 0$. Consequently, $a \wedge q_{\alpha+1}(x_1) \neq 0$, as was to be shown.

The inductive proof is finished. The assertion follows, because for α sufficiently big, $q_\alpha(x_i) = x_i$. $\square$

Proposition 2.6.5 (Lemma 5.7 in [54]) *Assume that A is rc-filtered and let τ be a regular and uncountable cardinal. If $\mathcal{I}$ is a set of ideals of A such that $\gamma(I, A) < \tau$ for all $I \in \mathcal{I}$, then there exists $\mathcal{J} \subseteq \mathcal{I}$ such that $|\mathcal{J}| < \tau$ and $\bigcap \mathcal{I} = \bigcap \mathcal{J}$.*

The following **proof** is a modification of the proof of Theorem 3 in [2] due to Bandlow. It is easier than the original argument. We derive a contradiction from the assumption that $\mathcal{J}$ does not exist. In that case we can find a sequence $(I_\alpha)_{\alpha < \tau}$ of ideals in $\mathcal{I}$ such that $\bigcap_{\alpha < \beta} I_\alpha \not\subseteq I_\beta$, for all $\beta < \tau$. Let $a_\beta \in A$ be such that

(1) $a_\beta \in \bigcap_{\alpha < \beta} I_\alpha$ and $a_\beta \notin I_\beta$.

Using $\gamma(I_\alpha, A) < \tau$ and the fact that A is rc-filtered, we can find a continuous chain $(B_\alpha)_{\alpha < \tau}$ of relatively complete subalgebras of A such that for all $\alpha < \tau$

(2) $a_\alpha \in B_{\alpha+1}$,

(3) $I_\alpha \cap B_{\alpha+1}$ generates I_α, and

(4) $|B_\alpha| < \tau$.

All jump sets that occur in the sequel refer to the filtration $(B_\alpha)_{\alpha < \tau}$ of $B = \bigcup_{\alpha < \tau} B_\alpha$. As above we denote the projection $B \to B_\alpha$ by q_α.

All $d(a_\alpha)$ being finite, we can apply the Δ-Lemma (cf. 0.0.2) to obtain $H \subseteq \tau$ of power τ such that $\{d(a_\alpha) : \alpha \in H\}$ is a Δ-system with kernel d_0, say. We fix an ordinal $\delta < \tau$ which is bigger than all elements of d_0.

The set $\{q_\delta(a_\alpha) : \alpha \in H\}$ is contained in B_δ, which, by (4), has power $< \tau$. It follows that we can find $G \subseteq H$ still of power τ such that

(5) $q_\delta(a_\alpha) = q_\delta(a_\beta)$, for all $\alpha, \beta \in G$.

Next we fix some $\alpha \in G$ such that $\alpha > \delta$. The interval $[\delta, \alpha + 1)$ having less than τ elements and $\{d(a_\beta) \setminus \delta : \beta \in G, \beta > \alpha\}$ consisting of τ pairwise disjoint sets, we can find some $\beta \in G$, $\beta > \alpha$ such that $d(a_\beta) \cap [\delta, \alpha + 1) = \emptyset$. From the definition of jump set it then follows that $q_{\alpha+1}(a_\beta) = q_\delta(a_\beta)$, which together with (5) yields

(6) $q_{\alpha+1}(a_\beta) = q_\delta(a_\alpha) \geq a_\alpha$.

To obtain the desired contradiction, we notice that, by (1) and (3), $\alpha < \beta$ implies $a_\beta \in I_\alpha$, hence $a_\beta \leq h$ for some $h \in I_\alpha \cap B_{\alpha+1}$. From $a_\beta \leq h \in B_{\alpha+1}$ and the definition of $q_{\alpha+1}$ we get $q_{\alpha+1}(a_\beta) \leq h$, hence $a_\alpha \leq h$, by (6).

But, $a_\alpha \leq h \in I_\alpha$ implies $a_\alpha \in I_\alpha$, which contradicts (1). $\square$

Corollary 2.6.6 *If A, τ and $\mathcal{I}$ are as in 2.6.5 above, then $\gamma(\bigcap \mathcal{I}, A) < \tau$.*

Proof.   Indeed, choose $\mathcal{J}$ according to the previous proposition and take a relatively complete subalgebra C of A such that C has cardinality $< \tau$ and contains a set of generators of all $J \in \mathcal{J}$. So $J \cap C$ generates J for all $J \in \mathcal{J}$.

To see that $\bigcap \mathcal{I}$ is also generated by its intersection with C, consider an arbitrary $a \in \bigcap \mathcal{I} = \bigcap \mathcal{J}$. Let $\bar{a}$ denote the minimal element of C above a. It will be sufficient to establish $\bar{a} \in \bigcap \mathcal{J}$.

For each $J \in \mathcal{J}$ we have $a \in J$, hence $a \leq b$ for some $b \in J \cap C$. But then $\bar{a} \leq b$, by minimality, which shows $\bar{a} \in J$, as desired. $\square$

We finish this section translating the last two statements into the language of filters.

Proposition 2.6.7 *Assume that A is rc-filtered and let τ be a regular and uncountable cardinal. If $\mathcal{F}$ is a set of filters of A such that $\chi(F, A) < \tau$ for all $F \in \mathcal{F}$, then there exists $\mathcal{G} \subseteq \mathcal{F}$ such that $|\mathcal{G}| < \tau$ and $\bigcap \mathcal{F} = \bigcap \mathcal{G}$. Moreover, $\chi(\bigcap \mathcal{F}, A) < \tau$.*

2.7   Computation of cardinal functions

Most results on cardinal functions that will be proved in this section hold not only for rc-filtered Boolean algebras but also for their subalgebras. In particular they generalize the corresponding results for subalgebras of free Boolean algebras. It will be convenient, though not very beautiful, to adopt the following mode of speech.

Definition 2.7.1 We say that a Boolean algebra is *subrcf* if it is a subalgebra of some rc-filtered Boolean algebra. □

Proposition 2.7.2 (Theorem 6 in [54]) *Every regular uncountable cardinal is a caliber of every subrcf algebra.*

Proof. We may assume that the algebra in question, A say, is itself rc-filtered. We argue by induction on $|A|$. There is nothing to prove if A is countable. An uncountable A has an open filtration $(A_\alpha)_{\alpha < \rho}$ consisting of subalgebras of strictly smaller power. The jump sets occuring below refer to this filtration.

Let X be a subset of A of regular uncountable cardinality, κ say. As jump sets are finite, we can apply the Δ-Lemma to $\{d(x) : x \in X\}$. It yields a subset Y of X, which has power κ and is such that $\{d(y) : y \in Y\}$ forms a Δ-system with kernel d_0, say. Let β be bigger than all elements of d_0. Then

$$d(y_1) \cap d(y_2) \cap [\beta, \rho) = \emptyset$$

for all $y_1 \neq y_2 \in Y$. By induction hypothesis, κ is a caliber of A_β. Consequently, there is a subset Z of Y still of power κ such that $\{q_\beta(z) : z \in Z\}$ has the finite intersection property. By 2.6.4, Z has the finite intersection property. □

For a Boolean algebra A and a cardinal number τ we introduce the notation

$$P(\tau, A) = \bigcap \{U \in UltA : \pi(U, A) < \tau\}.$$

If A is clear from the context, we only write $P(\tau)$. Being an intersection of (ultra)filters, $P(\tau)$ is a filter and we may form the quotient $A/P(\tau)$. It has small power, as shows

Proposition 2.7.3 (Theorem 11 in [54]) *If A is subrcf and τ regular uncountable, then $|A/P(\tau)| < \tau$.*

Proof. Take an rc-filtered B such that $A \leq B$. Denote the set $\{U \in UltA : \pi(U, A) < \tau\}$ by $\mathcal{U}$. For each $U \in \mathcal{U}$ we use 2.6.3 to fix a proper filter F_U of B for each $U \in \mathcal{U}$ such that $U \subseteq F_U$ and $\chi(F_U, B) < \tau$. Then 2.6.7 yields a subset $\mathcal{V}$ of $\mathcal{U}$ such that $|\mathcal{V}| < \tau$ and $G = \bigcap \{F_U : U \in \mathcal{V}\} = \bigcap \{F_U : U \in \mathcal{U}\}$. Using that B is rc-filtered, we can choose a subalgebra $C \leq_{rc} B$ of power $< \tau$ such that

$$(*) \qquad F_U \cap C \text{ generates } F_U \text{ for all } U \in \mathcal{V}.$$

CLAIM   *For each $a \in A$ there exists some $c \in C$ such that $a/G = c/G$.*

Indeed, as $a \in A \subseteq B \geq_{rc} C$, there exists a maximal element $c \in C$ below a. Clearly, $c/G \leq a/G$. To establish $a/G \leq c/G$ it is sufficient to check $(a - c)/G = 0/G$, i.e. $-(a - c) \in G$, i.e. $-(a - c) \in F_U$, for all $U \in \mathcal{V}$.

If $a \in U \subseteq F_U$, then, by $(*)$, there exists some $d \in F_U \cap C$ such that $d \leq a$. The choice of c yields $d \leq c$, hence $d \leq -(a - c) \in F_U$. If $a \notin U$, then $-a \in U \subseteq F_U$ and $-a \leq -(a - c) \in F_U$, again.

It follows from the claim that $A/(G \cap A)$ is isomorphic to a subalgebra of $C/(G \cap C)$. So $|A/(G \cap A)| \leq |C| < \tau$. But,

$$G \cap A = \bigcap \{F_U \cap A : U \in \mathcal{U}\} = \bigcap \{U : U \in \mathcal{U}\} = P(\tau),$$

and we are done. $\square$

For rc-filtered algebras the following result is Theorem 19 in [53]. The proof was more complicated then, because Šapirovskii's theorem was not yet available.

Corollary 2.7.4 *If A is subrcf, then the free Boolean algebra on τ generators embeds into A for each regular uncountable $\tau \leq |A|$. It follows that $\operatorname{ind}(A) = |A|$.*

Proof. From the above proposition we have that $|A/P(\tau)| < \tau \leq |A|$. It follows that there is some non-zero $a \in A$ which is disjoint from some element in $P(\tau)$. In other words, $\pi(U, A) \geq \tau$, for all $U \in \operatorname{Ult} A$ such that $a \in U$. By Šapirovskii's Theorem (10.16 in [34]), $A \restriction a$, hence A contains a copy of the desired free Boolean algebra.

As $\operatorname{ind}(A) = \sup\{|X| : \operatorname{Fr} X \text{ embeds into } A\}$ the second assertion follows at once. $\square$

Corollary 2.7.5 (Theorem 11 in [54]) *Let A be subrcf.*

(1) $\pi\chi(A) = \chi(A) = \pi(A) = |A|$

(2) *If $cf(|A|) > \omega$, then $\pi(U, A) = |A|$ for some $U \in \operatorname{Ult} A$.*

Proof. From the previous proof we know that for all regular $\tau \leq |A|$ there is some $U \in \operatorname{Ult} A$ such that $\pi(U, A) \geq \tau$. This immediately yields (2) for regular $|A|$ and also $|A| \leq \pi\chi(A)$.

As $\pi\chi(A) \leq \chi(A) \leq |A|$ and $\pi\chi(A) \leq \pi(A) \leq |A|$ are always true, we get (1) as well.

To prove (2) for singular $|A|$, we denote $cf(|A|)$ by κ and let $(\tau_\alpha)_{\alpha < \kappa}$ be an increasing sequence of regular cardinals such that $|A| = \sup \tau_\alpha$. As in the previous proof, we can take non-zero $a_\alpha \in A$ such that a_α is not contained in any ultrafilter U such that $\pi(U, A) < \tau_\alpha$. By 2.7.2, κ is a caliber of A. So we can take an ultrafilter $V \in \operatorname{Ult} A$ such that $S = \{\alpha : a_\alpha \in V\}$ has power κ and is, therefore, cofinal in κ. Clearly, $\pi(V, A) \geq \tau_\alpha$ for all $\alpha \in S$, hence $\pi(V, A) \geq |A|$. $\square$

The easiest example to show that, in general, neither independence nor π-character are attained is the weak product (cf. 2.5.2) $\Pi_{n<\omega}^w \operatorname{Fr} \omega_n$. It is rc-filtered, by 2.5.3. It has power $\aleph_\omega$ but no independent subset of power $\aleph_\omega$ (exercise or [42], 8.3) and no ultrafilter of character $= \pi$-character $\aleph_\omega$. Moreover, its number of ultrafilters is $\sum_{n<\omega} 2^{\aleph_n}$, which is $\aleph_\omega$ under GCH. Notice that for $|A|$ regular, the free subalgebra of power $|A|$ gives rise to the maximal possible number of $2^{|A|}$ ultrafilters.

We proceed with an assertion showing, in particular, that atomic rc-filtered Boolean algebras are countable.

Corollary 2.7.6 (Theorem 13 in [54]) *If a Boolean algebra is subrcf, then the ideal of its atomic elements is countable.*

Proof. Let A be subrcf. By 2.7.3, $A/P(\aleph_1)$ is countable. It will be sufficient to show that the canonical homomorphism $a \mapsto a/P(\aleph_1)$ becomes injective if it is restricted to the set of atomic elements. Otherwise, there were a non-zero atomic $a \in A$ such that $a/P(\aleph_1) = 0/P(\aleph_1)$, i.e. $a \wedge f = 0$ for some $f \in P(\aleph_1)$. Let U be a principal ultrafilter containing a. As $\pi(U, A) \leq \chi(U, A) = \aleph_0$, we have $f \in P(\aleph_1) \subseteq U$. But then $a \wedge f \in U \not\ni 0$, contradiction. $\square$

Clearly, all rc-filtered algebras are subrcf. So, the results obtained so far apply to them. If A is rc-filtered, then, by 2.6.2, $\pi(U, A) = \chi(U, A)$ for all $U \in Ult A$. Moreover, $P(\tau, A)$ is identical with

$$X(\tau, A) = \bigcap \{U \in Ult A : \chi(U, A) < \tau\}.$$

The following theorem lists what we get after we replace π by χ in the above results and two more assertions that are true for rc-filtered algebras only.

Theorem 2.7.7 *Assume that A is rc-filtered and let τ be regular and uncountable.*

(1) $|A/X(\tau)| < \tau$

(2) $\chi(X(\tau), A) < \tau$

(3) *For all $U \in Ult A$ it holds that $\chi(U, A) < \tau$ iff $X(\tau) \subseteq U$.*[7]

(4) *If $cf(|A|) > \omega$, then $\chi(U, A) = |A|$ for some $U \in Ult A$.*

Proof. (1) and (4) are immediate from the remarks above. (2) follows from 2.6.7. The implication from left to right in (3) is obvious. It remains to see that $X(\tau) \subseteq U$ implies $\chi(U, A) < \tau$. Using (1) and (2), we can take a subalgebra $B \leq A$ of power $< \tau$ such that

(*) B intersects every congruence class $a/X(\tau)$,   and

(**) $X(\tau) \cap B$ generates $X(\tau)$.

We show that $U \cap B$ generates U. Let $a \in U$ be given. Use (*) to choose $b \in B$ such that $b \in a/X(\tau)$, i.e. $b \wedge f = a \wedge f$ for some $f \in X(\tau) \subseteq U$. By (**), we can assume that f belongs to B. From $a \wedge f \in U$ we get $b \wedge f \in U \cap B$. So $b \wedge f \leq a$ proves what we want. $\square$

For an application in chapter 5.1.2 we draw the following corollary, originally due to Efimov.

Corollary 2.7.8 (cf. 3.8 in [35]) *If the projective Boolean algebra A has regular uncountable cardinality τ, then $A \restriction a \simeq Fr\,\tau$ for some $a \in A$.*

---

[7] In other terms: $\{U \in Ult A : \chi(U, A) < \tau\}$ is a closed subset of $Ult A$.

Proof. The theorem applies to A and τ, hence $|A/X(\tau)| < \tau = |A|$. It follows that there is some non-zero $a \in A$ such that all ultrafilters containing a have character τ. As $A \restriction a$ is projective (by 2.4.1), Ščepin's theorem 1.4.11 applies, hence $A \restriction a \simeq \operatorname{Fr} \tau$. $\square$

A slight modification will be useful in section 3.3.

Corollary 2.7.9 *If the projective Boolean algebra A has at least the regular uncountable cardinality τ, then $\operatorname{Fr} \tau$ is a retract of A.*

Proof. We know that A has an additive rc-skeleton. In 1.5.5 we have proved that it consists of projectively embedded subalgebras. This allows us to take some $B \leq_{proj} A$ such that $|B| = \tau$. By the previous corollary, $B \restriction b \simeq \operatorname{Fr} \tau$, for some $b \in B$. It follows that $\operatorname{Fr} \tau$ is a retract of B, which is a retract of A, by 1.5.6(2). $\square$

We end this section with an information that is adressed to those readers who know the cardinal functions involved. The proof is immediate from 2.2.7, 2.7.5 and known inequalities between cardinal functions (cf. the chart on p. 127 of [42]).

Information 2.7.10 *For subrcf algebras:*

$$\pi\chi = ind = \pi = Irr = t = s = \chi = hL = hd = Inc = h - cof = |\ \ |$$
$$\vee |$$
$$d$$
$$\vee |$$
$$Depth = Length = c = \aleph_0$$

The example of free Boolean algebras shows that $d = topological\ density$ can be strictly between $\aleph_0$ and $|A|$.

2.8 A characterization of the free Boolean algebra on $\aleph_1$ generators

We already know a characterization of free Boolean algebras among projective ones: by Ščepin's Theorem 1.4.11, all ultrafilters have to have the same character. In this section we prove

Theorem 2.8.1 ([6] and [51]) *A subrcf Boolean algebra of cardinality $\aleph_1$ is free iff it has a homogeneous Stone space, i.e. if for each pair of prime ideals[8] P, Q there exists an automorphism φ of the Boolean algebra such that $\varphi(P) = Q$.*

---

[8] Or ultrafilters if one desires.

In view of Ščepin's theorem mentioned above, it is natural to ask whether for the current theorem it will be sufficient that all ultrafilters have character $\aleph_1$. The answer is negative, by Engelking's example 2.3.4.

Notice that each subrcf algebra of power $\aleph_1$ is in fact a subalgebra of $\mathrm{Fr}\,\omega_1$ and vice versa. Indeed, assume $A \leq B$, where B is rc-filtered. Without loss, $|B| = \aleph_1$, so, by 2.2.7, B is projective hence a subalgebra of a free Boolean algebra.

Recall that a topological space is called *dyadic* if it is a continuous image of some generalized Cantor cube 2^κ. Using this notion, the topological dual of the theorem reads: *Any homogeneous, zero-dimensional dyadic space of weight $\aleph_1$ is homeomorphic to 2^{ω_1}*

The question whether each homogeneous, zero-dimensional dyadic space must be a Cantor cube was posed by B.A. Efimov. It had earlier been answered by V.V. Pašenkov, who constructed counter-examples. In our language his result reads as follows.

Information 2.8.2 *For each uncountable cardinal κ there is Boolean algebra of power 2^κ which has a homogeneous Stone space and is embeddable into a free Boolean algebra but does not have the Bockstein separation property. Such an algebra cannot be rc-filtered, let alone free.*

We refer the interested reader to Pašenkov's paper [43], which is short and very clearly written. As there may be some room between $\aleph_1$ and $2^{\aleph_1}$, there remain some open questions in the area.

Now we outline the **proof** of the theorem. From what we already know it will be easy to deduce the result, once we have established that A has the (BSP). Indeed, by 2.3.5, A will then itself be rc-filtered and, having power $\aleph_1$, even projective (2.2.7). Homogeneity further yields that all ultrafilters have the same character, which must be $\aleph_1$, by 2.7.5. It remains to apply Ščepin's abovementioned theorem.

How can homogeneity help to establish the (BSP)? The idea is to 'localize' it. We shall find a property of prime ideals such that a subrcf algebra has the (BSP) iff all of its prime ideals have this property. Then it will be sufficient to prove that each subrcf algebra of power $\aleph_1$ has at least one prime ideal with that particular property.

It may be easier to digest the following result if we first formulate it in topological terms. Recall that a subset U of a topological space X is called G_δ-*closed* if for each $p \notin U$ there is a G_δ-set G such that $p \in G$ and $G \cap U = \emptyset$.[9] The result then says that, under certain conditions on the space, every open G_δ-closed set is of type F_σ. For, say, dyadic spaces this[10] is well-known among topologists but not so easy to attribute. The most explicit early mentioning (for products of separable metric spaces) is probably in [12].

---

[9] That is to say that U is closed in the G_δ-topology on X, which is the topology with all originally G_δ-sets as a base.

[10] The more popular form asserts that closed G_δ-open sets are G_δ.

We now formulate and prove the assertion in Boolean algebraic terms and, to please the friends of generality, for arbitrary regular uncountable cardinals.

Lemma 2.8.3 *Let I be an ideal of a subrcf Boolean algebra A and let κ be an uncountable regular cardinal. If for every prime ideal $P \supseteq I$ there exists an ideal J_P such that $I \subseteq J_P \subseteq P$ and $\gamma(J_P, A) < \kappa$, then $\gamma(I, A) < \kappa$.*

Proof.   Take an rc-filtered $B \geq A$. For an ideal $H \subseteq A$ we denote by $\hat{H}$ the ideal of B generated by H, i.e.

$$\hat{H} = \{b \in B : b \leq h \text{ for some } h \in H\}.$$

Then, obviously, $H = \hat{H} \cap A$, and

$(*)$ $\gamma(\hat{H}, B) \leq \gamma(H, A)$

(just take the same set of generators). For the ideal I in question we now

CLAIM $\hat{I} = \bigcap\{\widehat{J_P} : I \subseteq P \text{ prime}\}$

Indeed, as $I \subseteq J_P$ implies $\hat{I} \subseteq \widehat{J_P}$ for all P, the inclusion from left to right is obvious.

To get the other inclusion we consider any $b \in B$, $b \notin \hat{I}$. The set $\{-b\} \cup I$ then generates a proper ideal of B, for, otherwise, $1 = -b \vee i$, i.e. $b \leq i$ for some $i \in I$. Extend $\{-b\} \cup I$ to a prime ideal Q of B. Then $P = Q \cap A \supseteq I$ is a prime ideal of A, for which there is some J_P. As $J_P \cup \{-b\} \subseteq Q$, we get $\widehat{J_P} \cup \{-b\} \subseteq Q$, hence $b \notin \widehat{J_P}$, as claimed.

From $(*)$ and corollary 2.6.6 we have that $\gamma(\hat{I}, B) < \kappa$. Let $I_0 \subseteq \hat{I}$ be a set of generators of power $< \kappa$. Witout loss we can assume that I_0 is closed under finite joins. Form I' by choosing an element $i' \in I$ above each element $i \in I_0 \subseteq \hat{I}$. To see that I' generates I, consider any $a \in I$. As $a \in \hat{I}$, there is $i \in I_0$ such that $a \leq i$. But then $a \leq i'$ and $i' \in I'$, as desired. As $|I'| \leq |I_0| < \kappa$, we are done.
□

Now all that is left in order to prove the theorem is the construction of *one* special prime ideal of the subrcf algebra A. In topological terms we need a point which whenever belonging to a regular closed set is a G_δ-inner point of that set.

Lemma 2.8.4 *Every subrcf Boolean algebra A of regular uncountable cardinality κ contains a prime ideal P such that for all regular ideals I such that $I \subseteq P$ there exists an ideal J such that $\gamma(J, A) < \kappa$ and $I \subseteq J \subseteq P$.*

Proof.   Fix an rc-filtered $B \geq A$ which has power κ. Let us enumerate B as $\{b_\alpha : \alpha < \kappa\}$ and denote the subsets $\{b_\alpha : \alpha < \delta\}$ by B_δ.

By induction on α, we define a mapping $f : B \to A$. Assume that $f(b_\beta)$ has been defined for all $\beta < \alpha$ in such a way that $f(B_\alpha) = \{f(b_\beta) : \beta < \alpha\}$ generates a proper ideal, i.e. $f(b_{\beta_1}) \vee f(b_{\beta_2}) \vee \ldots \vee f(b_{\beta_n}) < 1$, for any finite number of ordinals $\beta_1, \beta_2 \ldots, \beta_n < \alpha$.

If possible, the value $f(b_\alpha) \in A$ is chosen to be *above* b_α and such that $f(B_{\alpha+1}) = \{f(b_\alpha)\} \cup f(B_\alpha)$ still generates a proper ideal. If such a choice is impossible, we put $f(b_\alpha) = 0$.

Notice that $f(B) = \{f(b_\alpha) : \alpha < \kappa\} \subseteq A$ generates a proper ideal and extend this set to a prime ideal P of A. This P will be as desired.

To check this, we let I be a regular ideal of A such that $I \subseteq P$. We have to construct an ideal J such that $I \subseteq J \subseteq P$ and $\gamma(J, A) < \kappa$. It will be sufficient to find an ordinal $\delta < \kappa$ such that I is contained in the ideal generated by $f(B_\delta)$. For this ideal is contained in P because its generators are and it is generated by atmost $|f(B_\delta)| \leq |\delta| < \kappa$ elements.

By regularity, I can be written as M^d for a suitable set M of pairwise disjoint elements of A. By the ccc (holding in B hence in $A \leq B$), M is countable.

CLAIM *There exists an ordinal $\delta < \kappa$ such that the set B_δ is a relatively complete Boolean subalgebra of B, which contains M and is closed under f.*

Indeed, as B is rc-filtered, there exists a skeleton S of relatively complete subalgebras of B all of which are closed under f. As M is countable, we find a countable $S_0 \in S$ such that $M \subseteq S_0$. By parallel induction we then construct increasing sequences $(S_n)_{n<\omega}$ in S and $(\delta_n)_{n<\omega}$ in κ such that

$$M \subseteq S_0 \subseteq B_{\delta_0} \subseteq S_1 \subseteq B_{\delta_1} \ldots \subseteq S_n \subseteq B_{\delta_n} \subseteq S_{n+1} \ldots$$

This is easy, using the regularity of κ and the fact that S is absorbing. The ordinal $\delta = \sup \delta_n$ is as claimed, for $B_\delta = \bigcup_{n<\omega} B_{\delta_n} = \bigcup_{n<\omega} S_n \in S$.

To see that I is contained in the ideal, J say, generated by $f(B_\delta)$, we consider an arbitrary $a \in I = M^d$ and assume $a \notin J$, to get a contradiction. Notice first that $(*)$ $f(B_\delta) \cup \{-a\}$ generates a proper ideal, for, otherwise

$$-a \vee f(b_{\beta_1}) \vee \ldots \vee f(b_{\beta_n}) = 1, \text{ hence } a \leq f(b_{\beta_1}) \vee \ldots \vee f(b_{\beta_n}) \in J.$$

Let $\bar{a}$ denote the projection of a in B_δ. Then $-\bar{a} = b_\alpha$ for some $\alpha < \delta$, hence $a \leq \bar{a} \leq -b_\alpha$ and $(**)$ $b_\alpha \leq -a$.

From $(*)$ and $(**)$ it follows that it was possible to choose $f(b_\alpha) \geq b_\alpha$. As $f(b_\alpha) \in P \supseteq M^d$, we have $-f(b_\alpha) \notin M^d$, hence $-f(b_\alpha) \wedge m \neq 0$, for some $m \in M$.

As $-f(b_\alpha) \leq -b_\alpha = \bar{a}$, we get $\bar{a} \wedge m \neq 0$, for the same $m \in M$. It follows that $a \wedge m \neq 0$, for otherwise $a \leq \bar{a} - m < \bar{a}$ contradicting the minimality of $\bar{a}$ above a. But $a \wedge m \neq 0$ contradicts $a \in M^d$, which ends the proof. □

As explained above together the two lemmas prove the theorem.

We end this section with an **open problem.** *Is it true that each rc-filtered Boolean algebra with homogeneous Stone space must be free?* All the concrete examples of non-projective rc-filtered Boolean algebras we know[11] are non-homogeneous.

---

[11] The reader will know most of them soon.

2.9 κ-metrization

At the end of section 2.2 we mentioned that the dual spaces of rc-filtered Boolean algebras are called openly generated. Another name, which reflects the original definition of this class, is κ-metrizable space. We now want to give this definition and prove the easy direction of the equivalence.

This section will have a rather more topological flavour than the rest of this work. Let us fix a compact, zero-dimensional space X and consider its dual algebra $Clop(X)$ consisting of all clopen (=closed and open at the same time) subsets of X. Recall that a subset of X is called *regular closed* if it is the closure of some open set. We let $RC(X)$ denote the family of all regular closed subsets of X.

Definition 2.9.1 A κ-metric[12] on X is a function ρ that maps $X \times RC(X)$ to the non-negative reals and satisfies the following axioms.

$(K1)$ $x \in F$ iff $\rho(x, F) = 0$.

$(K2)$ $F \subseteq G$ implies $\rho(x, F) \geq \rho(x, G)$.

$(K3)$ The function $x \mapsto \rho(x, F)$ is continuous, for each fixed F.

$(K4)$ If $(F_\alpha)_{\alpha < \lambda}$ is a chain of regular closed sets, then

$$\rho(x, \overline{\bigcup_{\alpha < \lambda} F_\alpha}) = \inf\{\rho(x, F_\alpha) : \alpha < \lambda\}.$$

X is called κ-metrizable if there is a κ-metric on X. □

As to $(K4)$ the reader should notice that the closure $\overline{\bigcup_{\alpha < \lambda} F_\alpha}$ is automatically regular closed. The prototype of a κ-metric is the usual distance of a point from a closed set in metric spaces. But, and this is the reason for their introduction, κ-metrics exist also on non-metrizable spaces. Here is the promised characterization, due to Ščepin [54].

Theorem 2.9.2 *The space X is κ-metrizable iff $Clop(X)$ is rc-filtered.*

We only prove the implication from left to right. Let ρ be a fixed κ-metric on X. We have to distinguish subalgebras of $Clop(X)$ that make up an rc-skeleton. Every subalgebra $B \leq Clop(X)$ induces an equivalence relation $\equiv_B$ on X:

$$x \equiv_B y \quad \text{iff} \quad \forall b \in B \, (x \in b \Leftrightarrow y \in b).$$

The equivalence class of x is $x/_{\equiv_B} = \bigcap\{b \in B : x \in b\}$. Together with ρ, the subalgebra B induces a second equivalence relation $\cong_B$ on X:

$$x \cong_B y \quad \text{iff} \quad \forall b \in B \, \rho(x, b) = \rho(y, b).$$

It follows from (K1) that $\cong_B$ is stronger than $\equiv_B$. We call $B \leq Clop(X)$ *admissible* if the relations $\equiv_B$ and $\cong_B$ coincide.

---

[12] In the Russion literature regular closed sets are called *canonically* closed. That is where the κ comes from – it can be interpreted as a cyrillic 'k' as well.

Lemma 2.9.3 *The set of admissible subalgebras of $Clop(X)$ contains a skeleton.*

Proof. For every $a \in Clop(X)$ and every rational number r the open set

$$\{x \in X : \rho(x,a) > r\} = \bigcup_{n=1}^{\infty} \{x \in X : \rho(x,a) \geq r + 1/n\}$$

is a countable union of closed sets, by (K3). Therefore, we can fix a countable subset $F_r(a)$ of $Clop(X)$ such that

$$\{x \in X : \rho(x,a) > r\} = \bigcup F_r(a).$$

Define $F_\infty(a) = \bigcup \{F_r(a) : r$ rational $\}$. Then $F_\infty(a)$ is countable for all a. It will be sufficient to show that the fixed-point skeleton of F_∞ consists of admissible subalgebras only.

Let $B \leq Clop(X)$ be closed under F_∞ and assume $x \equiv_B y$. If $x \cong_B y$ were not true, we could find some $b \in B$ such that $\rho(x,b) \neq \rho(y,b)$. Take a rational r strictly between the two values, $\rho(x,b) < r < \rho(y,b)$, say. Then there is some $b' \in F_r(b) \subseteq B$ such that $x \notin b'$, but $y \in b'$. This contradicts $x \equiv_B y$. □

Our proof will be complete if we establish

Lemma 2.9.4 *If $B \leq Clop(X)$ is admissible, then $B \leq_{rc} Clop(X)$.*

Proof. We start with an assertion that has nothing to do with admissibility.

CLAIM   *If $x \cong_B y$, then $\rho(x, \overline{\bigcup M}) = \rho(y, \overline{\bigcup M})$ for all $M \subseteq B$.*

Notice that $\bigcup M$ is open, so its closure is regular closed and the application of ρ makes sense. To prove the claim, we use induction on the cardinality of M. If M is finite, then $\overline{\bigcup M} = \bigcup M = m \in B$, which makes the assertion trivial.

If M is infinite, we decompose it into a strictly ascending chain $M = \bigcup_{\alpha < \beta} M_\alpha$ of subsets of strictly smaller cardinalities.

Then $\overline{\bigcup M} = \overline{\bigcup_{\alpha < \beta} \overline{\bigcup M_\alpha}}$, hence

$$\rho(x, \overline{\bigcup M}) = \inf\{\rho(x, \overline{\bigcup M_\alpha}) : \alpha < \beta\}$$
$$= \inf\{\rho(y, \overline{\bigcup M_\alpha}) : \alpha < \beta\} = \rho(y, \overline{\bigcup M}),$$

by (K4).

Now let $a \in Clop(X)$ be given. We seek for $b \in B$ that is minimal above a. In other words, denoting $\{b \in B : a \wedge b = 0\}$ by M, we need some b such that

$$a \leq b \in B \quad \text{and} \quad b \cap \bigcup M = \emptyset.$$

Fix any $x \in a$. Then, as $a \cap \bigcup M = \emptyset$, $x \notin \overline{\bigcup M}$, hence $\rho(x, \overline{\bigcup M}) > 0$, by (K1). From admissibility and the claim it follows that the whole $\equiv_B$-class of x is disjoint from $\overline{\bigcup M}$, i.e.

$$\bigcap \{b \in B : x \in b\} \cap \overline{\bigcup M} = \emptyset.$$

Compactness yields some $b_x \in B$ such that $x \in b_x$ and $b_x \cap \bigcup M = \emptyset$. Again by compactness, finitely many b_x suffice to cover a. Their union is the sought-for b. $\square$

The proof of the other direction of the theorem is rather complicated and far from the rest of our presentation of the theory; we refer the interested reader to [54].

2.10   Answers to some questions of Ščepin

Recall that a topological space is said to be perfectly normal if all open subsets are of type F_σ. A well-known topological theorem ([13], 4.2.B, attributed to Šneider) asserts that a compact Hausdorff space X is metrizable if and only if $X \times X$ is perfectly normal. This result led Ščepin to ask (Question 9 in [55]) whether X must admit a κ-metric[13], if $X \times X$ is perfectly κ-normal. In our language: If $A \otimes A$ has the Bockstein separation property, does it follow that the Boolean algebra A is rc-filtered?

Using $\Diamond$, a first counterexample was constructed by A.V. Ivanov in [26]. It is quite complicated and has other (perhaps more interesting) properties.

The example below was constructed by Fuchino, Koppelberg, and Takahashi [18] for other purposes. Sakaé Fuchino then noticed that it answers two of Ščepin's questions, namely 9 and 3 in [55].

Proposition 2.10.1 (Fuchino [16]) *There exists a Boolean algebra A of cardinality $\aleph_1$ such that*

(1) *A is not rc-filtered;*

(2) *$A \otimes A$ has the Bockstein separation property;*

(3) *the family of relatively complete subalgebras of A is absorbing.*

Proof.   The desired Boolean algebra A will be constructed as a union of a continuous chain $(A_\alpha)_{\alpha < \omega_1}$ of countable free Boolean algebras. As a parameter we use a set $S \subseteq \omega_1$ which consists of limit ordinals only and is such that S and $\omega_1 \setminus S$ are both stationary[14]. The chain will satisfy

(i) $A_{\alpha+1} \leq_{free} A_{\beta+1}$ for all $\alpha < \beta$,   and   (ii) $A_\alpha \leq_{rc} A_{\alpha+1}$ iff $\alpha \notin S$.

We START with a copy of $\mathrm{Fr}\,\omega$. At LIMIT STEPS there is no choice, we must take unions. The SUCCESSOR STEP is easy if $\alpha \in S$: just take $A_{\alpha+1} = A_\alpha \otimes \mathrm{Fr}\,\omega$.

Assume then that $\alpha \notin S$. Take a strictly increasing sequence $(\alpha_n)_{n < \omega}$ of successor ordinals such that $\alpha = \sup \alpha_n$. By inductive assumption, $A_{\alpha_n} \leq_{free} A_{\alpha_{n+1}}$, which allows us to take pairwise disjoint countably infinite sets $X_0, X_1, \ldots$

---

[13] See the section 2.9.

[14] i.e. intersect all closed and unbounded subsets of ω_1

of free generators of $A_\alpha = \bigcup_{n<\omega} A_{\alpha_n}$ such that $A_{\alpha_n} = \langle X_0 \cup \ldots \cup X_n \rangle$. Choose any $x_n \in X_n$ and put

$$B_n = \langle x_0, \ldots, x_n \rangle, \quad C_n = \langle X_0 \cup \ldots \cup X_n \setminus \{x_0, \ldots, x_n\}\rangle,$$

$$B = \bigcup_{n<\omega} B_n, \quad \text{and } C = \bigcup_{n<\omega} C_n.$$

So, slightly abusing notation,

$$A_{\alpha_n} = B_n \otimes C_n, \quad \text{and} \quad A_\alpha = B \otimes C.$$

Clearly, there is an embedding $B \leq B' \simeq \mathrm{Fr}\,\omega$ which is not relatively complete. We put

$$A_{\alpha+1} = B' \otimes C.$$

Then $A_\alpha = B \otimes C \leq B' \otimes C = A_{\alpha+1}$ is not relatively complete. As all B_n are finite, Sirota's Lemma 1.4.10 yields, $B_n \leq_{free} B'$, hence $A_{\alpha_n} = B_n \otimes C_n \leq_{free} B' \otimes C = A_{\alpha+1}$, which implies $A_{\gamma+1} \leq_{free} A_{\alpha+1}$ for all $\gamma < \alpha$, by transitivity. Indeed, $A_{\gamma+1} \leq_{free} A_{\alpha_n} \leq_{free} A_{\alpha+1}$, for some n. This completes the construction of the chain.

Now we check that the algebra $A = \bigcup_{\alpha<\omega_1} A_\alpha$ has the desired properties.

(1) It is not rc-filtered, because, S being stationary, it has too many non-rc subalgebras.

(2) Notice that $(A_\alpha \otimes A_\alpha)_{\alpha<\omega_1}$ is a filtration of $A \otimes A$ that consists of countable subalgebras and is such that

$$\{\alpha : A_\alpha \otimes A_\alpha \leq_{rc} A \otimes A\} = \{\alpha : A_\alpha \leq_{rc} A\} = \omega_1 \setminus S$$

is stationary. The Bockstein separation property follows from the more general lemma below.

(3) As $\omega_1 \setminus S$ is also stationary, hence unbounded, A has an absorbing family of relatively complete subalgebras. $\square$

Lemma 2.10.2 *Assume that $(B_\alpha)_{\alpha<\omega_1}$ is a filtration of B with all B_α being countable subalgebras and such that $T = \{\alpha < \omega_1 : B_\alpha \leq_{rc} B\}$ is stationary. Then B satisfies the ccc and has the Bockstein separation property.*

The proof is adapted from that of Solovay and Tennenbaum's theorem 6.3 in [58]. For each α we define $q_\alpha : B \to B_\alpha$ by

$$q_\alpha(b) = \begin{cases} \min\{c \in B_\alpha : c \geq b\}, & \text{if this minimum exists,} \\ 1, & \text{otherwise.} \end{cases}$$

Assume that $W \subseteq B$ is infinite and consists of pairwise disjoint elements. We prove that W is countable. As each B_α is countable, each $q_\alpha(W) \subseteq B_\alpha$ is also countable and we can fix $\eta(\alpha) > \alpha$ such that $q_\alpha(W) = q_\alpha(W \cap B_{\eta(\alpha)})$. The set

$$G = \{\gamma < \omega_1 : \gamma \text{ is a limit ordinal and } \alpha < \gamma \text{ implies } \eta(\alpha) < \gamma\}$$

is easily seen to be closed and unbounded. Take $\gamma \in T \cap G$ such that $W \cap B_\gamma$ is infinite. In order to prove

CLAIM 1. $W \subseteq W \cap B_\gamma$

we take an arbitrary $w \in W$. Then $\gamma \in T$ implies $q_\gamma(w) = \min\{c \in B_\gamma : c \geq w\}$, which is less than 1 because $B_\gamma \cap W$ is infinite. Moreover, $q_\gamma(w) \in B_\gamma = \bigcup_{\alpha < \gamma} B_\alpha$, hence $q_\gamma(w) \in B_\alpha$ for some $\alpha < \gamma$.

It follows that $q_\gamma(w)$ is also the minimal element of B_α above w. So, $q_\alpha(w) = q_\gamma(w) < 1$. As $\gamma \in G$, we have $q_\alpha(w) = q_\alpha(w')$ for some $w' \in W \cap B_{\eta(\alpha)} \subseteq W \cap B_\gamma$.

From $w \wedge w' = 0$ and the definition of q_γ we get $q_\gamma(w) \wedge w' = 0$, hence $q_\alpha(w') \wedge w' = 0$. As $q_\alpha(w') \geq w'$, we get $w' \wedge w' = 0$, a contradiction.

We have proved $w \wedge w' \neq 0$. By disjointness of W, we have $w = w' \in W \cap B_\gamma$, which proves the claim. It yields the ccc at once.

To prove the Bockstein separation property, we let $I \subseteq B$ be a regular ideal. Then $I = I^{dd} = W^d$ for some $W \subseteq B$ consisting of pairwise disjoint elements (cf. 1.2.15). As we have just proved, W is countable. So $W \subseteq B_\alpha \leq_{rc} B$ for some $\alpha \in T$.

CLAIM 2. *The countable set $I \cap B_\alpha$ generates I as an ideal.*

Consider $i \in I = W^d$. It is sufficient to establish $q_\alpha(i) \in I$, i.e. $q_\alpha(i) \wedge w = 0$ for all $w \in W$. But $i \in I = W^d$ implies $i \wedge w = 0$, hence $i \leq -w \in B_\alpha$ and $i \leq q_\alpha(i) \leq -w$. □

Let us mention here to more answers to questions from [55].

Information 2.10.3 ([52]) *Under GCH there is an rc-filtered Boolean algebra without contably infinite homomorphic images.*

This algebra cannot have a free quotient (answer to question 4) and cannot embed into an exponential[15] (answer to question 5).

---

[15] To be defined in section 3.3 below.

Chapter 3

Functors

In this chapter we are concerned with some constructions of Boolean algebras that are in fact covariant functors. The first section begins with some general explanations that are not too rigorous but, hopefully, sufficient to understand what happens in concrete situations. Also in the first section there is Ivanov's general theorem saying that some functors turn rc-filtered Boolean algebras into projective Boolean algebras. In the remaining sections we are concerned with concrete functors. In section 2 we prove that superextensions satisfy Ivanov's conditions, so A is rc-filtered iff λA is projective. In sections 3 and 4 we use the functors exp and SP to produce examples of rc-filtered Boolean algebras that are not projective.

3.1 Generalities

For us a *functor* is a procedure F which assigns a Boolean algebra $F(A)$ to every Boolean algebra A and a homomorphism $F(\varphi) : F(A) \to F(B)$ to every homomorphism $\varphi : A \to B$ in such a way that the following natural conditions are satisfied

$$F(\varphi \circ \psi) = F(\varphi) \circ F(\psi) \quad \text{and} \quad F(id_A) = id_{F(A)}.$$

Let F be a functor. We say that F *preserves* a property of Boolean algebras resp. homomorphisms if $F(A)$ resp. $F(\varphi)$ has this property, whenever A resp. φ has this property. We say that F *preserves cardinalities* if $|F(A)| \leq |A|$ for all Boolean algebras A. Recall that, by definition, $|X|$ is always infinite.

Suppose that F preserves injectivity of homomorphisms and let $A \leq B$ be an embedding. We can consider the homomorphism $in : A \to B$ sending $a \in A$ to itself. Then $F(in) : F(A) \to F(B)$ is an injective homomorphism. We usually identify $F(A)$ with its image under $F(in)$ and write $F(A) \leq F(B)$.

An injectivity preserving functor F is said to *preserve relative completeness* of embeddings if $A \leq_{rc} B$ implies $F(A) \leq_{rc} F(B)$. We say that F *reflects* relative completeness if $A \leq B$ and $F(A) \leq_{rc} F(B)$ implies $A \leq_{rc} B$. Analogously with other properties of embeddings.

A functor F will be called *continuous* if it preserves injectivity[1] and if for all Boolean algebras A and all $x \in F(A)$ there is a finite subalgebra $A_x \leq A$ such that $x \in F(A_x)$. It is clear that a continuous functor preserves unions of chains, i.e. $F(\bigcup_{\alpha < \tau} B_\alpha) = \bigcup_{\alpha < \tau} F(B_\alpha)$, whenever $(B_\alpha)_{\alpha < \tau}$ is a chain. As an exercise the reader may prove that continuity and preservation of unions of chains are actually equivalent (use induction on $|A|$).

It is obvious that a continuous functor preserves cardinalities if it preserves finiteness. For later reference the following observations get a number.

Proposition 3.1.1 *Let F be a continuous functor that preserves cardinalities.*

(1) *For all $B \leq F(A)$ there is some $C \leq A$ such that $|C| \leq |B|$ and $B \leq F(C)$.*

(2) *If S is a skeleton of $F(A)$, then $T = \{T \leq A : F(T) \in S\}$ is a skeleton of A to be called the preimage of S under F.*

(3) *If, in addition, F preserves relative completeness of embeddings, then it preserves the class of rc-filtered Boolean algebras.*

(4) *If, in addition, F reflects relative completeness of embeddings, then it reflects the class of rc-filtered Boolean algebras, i.e. A is rc-filtered if $F(A)$ is rc-filtered.*

Proof. (1) is obvious.

(2) As F preserves unions of chains, it is clear that T is closed. To check absorption take any $C \leq A$. Put $C_0 = C$ and take $S_0 \in S$ such that $F(C_0) \leq S_0$ and $|S_0| \leq |F(C_0)| \leq |C|$. Continue by induction: use (1) to find $C_{n+1} \leq A$ such that $C_n \leq C_{n+1}$, $S_n \leq F(C_{n+1})$, and $|C_{n+1}| \leq |S_n| \leq |C|$. Then take $S_{n+1} \in S$ such that $F(C_{n+1}) \leq S_{n+1}$ and $|S_{n+1}| \leq |F(C_{n+1})| \leq |C|$. The union $\bigcup_{n < \omega} C_n$ contains C and has power $|C|$. Moreover, it belongs to T, for $F(\bigcup_{n < \omega} C_n) = \bigcup_{n < \omega} F(C_n) = \bigcup_{n < \omega} S_n \in S$.

As to (3), we prove by induction on $|A|$ that $F(A)$ is rc-filtered provided that A is rc-filtered. For $|A| = \aleph_0$, $|F(A)| = \aleph_0$ and $F(A)$ is rc-filtered. For uncountable A, we take an rc-filtration $(A_\alpha)_{\alpha < \rho}$ of A such that all A_α have power $< |A|$. By, 2.3.1 and induction hypothesis, all $F(A_\alpha)$ are rc-filtered. Continuity and preservation of relative completeness yield that $(F(A_\alpha))_{\alpha < \rho}$ is an rc-filtration of $F(A)$. It remains to apply 2.2.8.

(4) Let S be a skeleton for $F(A)$. By (2), its preimage is a skeleton for A, which is rc, because relative completeness is reflected. $\square$

As we see in section 3.3, a functor as in 3.1.1(3) does not necessarily preserve projectivity. For, even if F preserves the cardinality of Boolean algebras, it

---

[1] This is for simplicity only.

need not preserve weights of embeddings. But if we strengthen the hypothesis of rc-preservation, then projectivity is not only preserved but even acquired.

Let us say that F *strongly preserves relative completeness* if for each finite collection $B_1, \ldots, B_n$ of relatively complete subalgebras of A the subalgebra of $F(A)$ generated by $F(B_1) \cup \ldots \cup F(B_n)$ is relatively complete.

In the following we use boldface capitals to denote finite collections of subalgebras. If $\boldsymbol{B} = \{B_1, \ldots, B_n\}$, we write $\boldsymbol{B} \leq_{rc} A$ to mean that $B_i \leq_{rc} A$ for all i. Another convenient notation is $F(\boldsymbol{B})$ for $\langle F(B_1) \cup \ldots \cup F(B_n) \rangle_{F(A)}$.

Using these abbreviations, strong preservation of relative completeness can be expressed by the implication

$$\boldsymbol{B} \leq_{rc} A \implies F(\boldsymbol{B}) \leq_{rc} F(A).$$

The following result is due to Ivanov [28]. The proof given below seems to be new.

Theorem 3.1.2 *Let F be a continuous functor that preserves cardinalities. If F strongly preserves relative completeness of embeddings, then $F(A)$ is projective whenever A is rc-filtered.*

Proof. If A is countable, then $F(A)$ is also countable, hence projective. For uncountable A we prove by induction on $|A|$ that there is a limit ordinal ρ and a sequence $(\boldsymbol{B}_\alpha)_{\alpha < \rho}$ of finite sets of subalgebras of A such that

(o) $\boldsymbol{B}_\alpha \leq_{rc} A$,

(i) $\boldsymbol{B}_0$ consists of countable subalgebras of A,

(ii) $F(\boldsymbol{B}_{\alpha+1})$ is countably generated over $F(\boldsymbol{B}_\alpha)$, and

(iii) $(F(\boldsymbol{B}_\alpha))_{\alpha < \rho}$ is a filtration of $F(A)$.

Noticing that, by (o) and strong rc-preservation, $F(\boldsymbol{B}_\alpha) \leq_{rc} F(A)$, we may then apply 1.3.2(6) to conclude that $F(A)$ is projective.

Assume first that $|A| = \aleph_1$. Using that A is rc-filtered we find a well-ordered rc-filtration $(A_\alpha)_{\alpha < \omega_1}$ of A consisting of countable subalgebras. Then the sequence $(\{A_\alpha\})_{\alpha < \omega_1}$ is, obviously, as desired.

Suppose now that $|A| > \aleph_1$. Again we take an rc-filtration $(A_\alpha)_{\alpha < \kappa}$ of A such that $|A_\alpha| < |A|$ for all α. By 2.3.1 all $A_\alpha \leq_{rc} A$ are rc-filtered. It follows that the induction hypothesis applies to each A_α. Fix sequences $(\boldsymbol{B}^\alpha_\beta)_{\beta < \rho_\alpha}$ satisfying conditions $(o) - (iii)$ with respect to A_α. Notice that $\boldsymbol{B}^\beta_\alpha \leq_{rc} A$, by transitivity of relative completeness.

The sequence $(\{A_\alpha\})_{\alpha < \kappa}$ satisfies condition (o) and, by continuity of F, also (iii). We are going to refine this sequence in front of $\{A_0\}$ and between all $\{A_\alpha\}$ and $\{A_{\alpha+1}\}$ to make it satisfy conditions (i) and (ii) as well.

IN FRONT OF $\{A_0\}$: If A_0 is countable, we do nothing because (i) is already true. Otherwise, we put $(\boldsymbol{B}^0_\beta)_{\beta < \rho_0}$ in front of $\{A_0\}$. As $\boldsymbol{B}^0_0$ consists of countable algebras, condition (i) is now satisfied. Clearly, (o) and (iii) are still true.

BETWEEN $\{A_\alpha\}$ and $\{A_{\alpha+1}\}$: If $F(A_{\alpha+1})$ happens to be countably generated over $F(A_\alpha)$, we do nothing. Otherwise, we insert the sequence

$$(\boldsymbol{B}_\beta^{\alpha+1} \cup \{A_\alpha\})_{\beta < \rho_{\alpha+1}}$$

between $\{A_\alpha\}$ and $\{A_{\alpha+1}\}$ (keeping the two). It is quite obvious (use the continuity of F and $\langle \, \cdot \, \rangle$ and the fact that $\rho_{\alpha+1}$ is a limit ordinal) that

$$(F(\boldsymbol{B}_\beta^{\alpha+1} \cup \{A_\alpha\}))_{\beta < \rho_{\alpha+1}}$$

is a filtration of $F(A_{\alpha+1})$. Consequently, condition (*iii*) still holds. We have gained that condition (*ii*) is now satisfied between $\{A_\alpha\}$ and $\{A_{\alpha+1}\}$. Indeed, $F(\boldsymbol{B}_0^{\alpha+1} \cup \{A_\alpha\})$ is generated over $F(A_\alpha)$ by the countable set $F(\boldsymbol{B}_0^{\alpha+1})$. For the other successor steps we notice that the same countable set generating $F(\boldsymbol{B}_{\beta+1}^{\alpha+1})$ over $F(\boldsymbol{B}_\beta^{\alpha+1})$ also generates $F(\boldsymbol{B}_{\beta+1}^{\alpha+1} \cup \{A_\alpha\})$ over $F(\boldsymbol{B}_\beta^{\alpha+1} \cup \{A_\alpha\})$.

After all these insertions are made, the resulting sequence is as desired. $\square$

3.2   Superextensions

Here we use 3.1.2 to prove that a Boolean algebra is rc-filtered iff its superextension is projective. This result is due to Ivanov [27] and came as a big surprise, when it was first obtained: Historically, the special case 3.2.6 came first and the general theorem 3.1.2 was destilled from it later.

Following de Groot, topologists define the superextension of a topological space as the space of all maximal linked (i.e. having pairwise non-zero intersections) systems of its closed subsets. If the space is Boolean, it is sufficient to consider maximal linked systems of clopen sets. This shows that one can make the same construction knowing only the Boolean algebra of all clopen subsets. What we do below will be a mixture of algebra and topology. Instead of working with maximal linked systems, we use their characteristic functions. For a while 2 will mean the discrete space $\{0,1\}$ and the two-element Boolean algebra at almost the same time. This will, hopefully, not cause any trouble.

Definition 3.2.1 For a Boolean algebra A we denote by $L(A)$ the subspace of 2^A consisting of those $p : A \to 2$ that are order-preserving, i.e. $a \leq b$ implies $p(a) \leq p(b)$, and respect complements, i.e. $p(-a) = -p(a)$. By λA we denote $Clop(L(A))$, the algebra of closed and open subsets of $L(A)$.

Let $\varphi : A \to B$ be a homomorphism. If $p : B \to 2$ belongs to $L(B)$, then $p \circ \varphi$ is easily seen to belong to $L(A)$. Moreover, by defining $L(\varphi)(p) = p \circ \varphi$ we get a continuous mapping $L(B) \to L(A)$. Its Stone dual is declared to be $\lambda\varphi$. $\square$

The procedure λ is a functor, called *superextension*[2]. The necessary verifications are left to the reader, who is also asked to check that $L(A)$ is a closed subspace

---

[2]Obviously, all homomorphisms $A \to 2$ belong to $L(A)$. Consequently, $L(A)$ is an *extension* of the Stone space of A. It will not become clear from our treatment what is so *super* about it.

of 2^A. As λA is, by definition, an algebra of sets, it will be natural to use the set-theoretic notations for its operations. That will distinguish them from the operations in A.

For $a \in A$ we put $N(a) = \{p \in L(A) : p(a) = 1\}$. We write $N(\vec{a}) = N(a_1, \ldots, a_n)$ as an abbreviation for

$$N(a_1) \cap \ldots \cap N(a_n) = \{p \in L(A) : p(a_1) = \ldots = p(a_n) = 1\}.$$

It is clear that all $N(\vec{a})$ belong to λA. Moreover,

Lemma 3.2.2 *Every non-zero element of λA is a finite union of $N(\vec{a})$'s.*

Proof. By definition (of the product topology of 2^A) each element of λA is a finite union of intersections of $N(a)$'s and complements of $N(a)$'s. But, as all $p \in L(A)$ respect complements, $L(A) \setminus N(a) = N(-a)$. So finite intersections of $N(a)$'s suffice and that is what the lemma says. $\square$

Next we list the necessary combinatorical properties of N.

Lemma 3.2.3 *Assume $a_1, \ldots, a_n, b_1 \ldots, b_m \in A$.*

(1) *If each b_j is above some a_i, then $N(\vec{a}) \subseteq N(\vec{b})$.*

(2) $N(\vec{a}) \neq \emptyset$ *iff for all i, j $a_i \wedge a_j \neq 0$.*

Proof. (1) follows from the monotonicity of the $p \in L(A)$. To prove (2) we take $p \in N(\vec{a})$ and assume, by contradiction that $a_i \wedge a_j = 0$. Then $a_i \leq -a_j$, hence $1 = p(a_i) \leq p(-a_j) = -p(a_j) = -1 = 0$, contradiction. Assume, the other way round, that $a_i \wedge a_j \neq 0$ for all i, j. Use Zorn's Lemma to get a maximal subset P of A that contains $\vec{a}$ and is such that $b \wedge c \neq 0$ for all $b, c \in P$. Define $p : A \to 2$ by setting $p(a) = 1$ iff $a \in P$. It is sufficient to show that p is monotone and respects complements. The latter is immediate from the following

CLAIM *For all $a \in A$, exactly one of $a, -a$ belongs to P.*

It is clear that they cannot both belong to P. If neither would, then, by maximality of P, $a \wedge b = 0$ and $-a \wedge c = 0$ for some $b, c \in P$. But then $b \wedge c = (a \vee -a) \wedge (b \wedge c) = 0$, contradiction.

Monotonicity also follows from the claim. Indeed, assume $a \leq b$, $p(a) = 1$ but $p(b) = 0$. Then $a \in P$, $-b \in P$ and $a \wedge -b = 0$, contradiction. $\square$

Lemma 3.2.4 *The mapping sending $N(a) \in \lambda A$ to $a \in A$ extends to a homomorphism $\pi_A : \lambda A \to A$. Its right inverse $N : A \to \lambda A$ is order-preserving and respects disjointness.*

Proof. To see this we apply Sikorskis Extension Criterion 0.0.1. Assuming $N(a_1) \cap \ldots \cap N(a_n) \subseteq N(b_1) \cup \ldots \cup N(b_m)$ we have to show $a_1 \wedge \ldots \wedge a_n \leq b_1 \vee \ldots \vee b_m$. If that were not true, then all pairwise intersections of $a_1, \ldots, a_n, -b_1 \ldots, -b_m$ were non-zero. Consequently, $N(\vec{a}, -\vec{b}) \neq \emptyset$, by

3.2.3(2). But every element in $N(\vec{a}, -\vec{b})$ contradicts $N(a_1) \cap \ldots \cap N(a_n) \subseteq N(b_1) \cup \ldots \cup N(b_m)$. So, π_A exists.

The mapping $N : A \to \lambda A$ is, by definition, a right inverse of π_A. By 3.2.3, N respects disjointness and the partial order. $\square$

Let us now consider a homomorphism $\varphi : A \to B$ and figure out how $\lambda\varphi$ acts on the generators of λA. To stress where we are, we write N_A and N_B.

$$\begin{aligned}
\lambda\varphi(N_A(a)) &= L(\varphi)^{-1}(N(a)) \\
&= \{p \in L(B) : L(\varphi)(p)(\underline{a}) = p(\varphi(a)) = 1\} = N_B(\varphi(a)).
\end{aligned}$$

It follows that $\lambda\varphi(N_A(a_1, \ldots, a_n)) = N_B(\varphi(a_1), \ldots \varphi(a_n))$. Using this equation together with 3.2.2 and 3.2.3(2), we get that λ preserves the injectivity of homomorphisms. In particular, for $A \leq B$, λA is canonically isomorphic to the subalgebra of λB generated by all $N(a)$ for $a \in A$. It is also clear that λ is continuous.

Proposition 3.2.5 *The functor λ reflects and strongly preserves relative completeness of embeddings.*

Proof. We start with REFLECTION. Assume $A \leq B$ and $\lambda A \leq_{rc} \lambda B$. Let any $b \in B$ be given. We look for a maximal $a \in A$ below b. There is a maximal element of λA below $N(b)$, which we can write as $\bigcup_{i=1}^n N(\vec{a}_i)$, assuming that all $N(\vec{a}_i)$ are non-empty. For each i we can fix some j_i such that $a_{ij_i} \leq b$. For, otherwise $N(\vec{a}_i) \cap N(-b) = N(\vec{a}_i, -b)$ were non-empty, by 3.2.3(2), contradicting $N(\vec{a}_i) \subseteq N(b)$. Put $a = \bigvee_{i=1}^n a_{ij_i}$. Clearly, $a \in A$ and $a \leq b$. If a were not maximal, we could find some $c \in A$ such that $0 < c \leq b$ and $c \wedge a = 0$. But then $N(c) \in \lambda A$, $\emptyset \neq N(c) \subseteq N(b)$, and, for all i, $N(c) \cap N(\vec{a}_i) = \emptyset$, contradicting the maximality of $\bigcup_{i=1}^n N(\vec{a}_i)$ below $N(b)$. Reflection is proved.

To establish STRONG PRESERVATION, we consider $A_1, \ldots, A_n \leq_{rc} B$. Observe that $\lambda A = \langle \lambda A_1 \cup \ldots \cup \lambda A_n \rangle_{\lambda B}$ is generated by all $N(a)$'s with $a \in A_1 \cup \ldots \cup A_n$. Consequently, its non-zero elements can be written as finite unions of $N(\vec{a})$'s, where all $\vec{a}$'s belong to $\bigcup_{i=1}^n A_i$ (cf. 3.2.2).

By 1.2.3 and 3.2.2 it is sufficient to indicate a minimal element in λA above an arbitrarily given non-empty $N(b_1, \ldots, b_m) \in \lambda B$. For $j = 1, \ldots n$ take a minimal element $a_i^j \in A_j$ above b_i. Then $N(b_1, \ldots, b_m) \subseteq N(a_1^1, \ldots, a_m^1, \ldots, a_1^n, \ldots, a_m^n)$, by 3.2.3(1). If $N(\vec{a})$ were not minimal above $N(\vec{b})$, then we could find some $N(\vec{c}) \in \lambda A$, i.e. $\vec{c} \in A_1 \cup \ldots \cup A_n$, such that $N(\vec{c}, \vec{b}) = N(\vec{c}) \cap N(\vec{b}) = \emptyset$ but $N(\vec{c}, \vec{a}) = N(\vec{c}) \cap N(\vec{a}) \neq \emptyset$. From 3.2.3(2) we get (indices without loss equal 1) that $b_1 \wedge c_1 = 0$. Assume that $c_1 \in A_j$. Then $a_1^j \wedge c_1 = 0$, because a_1^j was minimal above b_1 in A_j. From 3.2.3(2) it now follows that $N(\vec{c}, \vec{a}) = \emptyset$, contradiction. $\square$

Now we have everything ready to prove Ivanov's

Theorem 3.2.6 ([27]) *For a Boolean algebra A the following assertions are equivalent.*

(1) *A is rc-filtered.* (2) *λA is projective.* (3) *λA is rc-filtered.*

Proof. (1) $\Rightarrow$ (2) follows from 3.1.2 and the lemmas above. (2) $\Rightarrow$ (3) is trivial. (3) $\Rightarrow$ (1) follows from lemmas 3.2.5 and 3.1.1(4) or, alternatively, from 2.4.5 and lemma 3.2.4 above. $\square$

We are now in a position to prove a result that was annonced in section 2.4.

Proposition 3.2.7 *All homomorphisms onto rc-filtered Boolean algebras have right inverses that respect both, the partial order and disjointness.*

Proof. Consider a surjective homomorphism $\varphi : A \to B$, where B is rc-filtered. We have to construct a mapping $\varepsilon : B \to A$ which is monotone and preserves disjointness and satisfies $\varphi \circ \varepsilon = id_B$.

φ gives rise to the following diagram, in which π_A and π_B denote the homomorphisms introduced in lemma 3.2.4.

$$
\begin{array}{ccc}
A & \xrightarrow{\ \varphi\ } & B \\[2pt]
\Big\uparrow{\scriptstyle \pi_A} & & \Big\uparrow{\scriptstyle \pi_B} \\[2pt]
\lambda A & \xrightarrow{\ \lambda\varphi\ } & \lambda B
\end{array}
$$

As all mappings are homomorphisms, to prove commutativity, it is sufficient to consider generators of λA, i.e. elements of the form $N_A(a)$:

$$
\pi_B(\lambda\varphi(N_A(a))) = \pi_B(N_B(\varphi(a))) = \varphi(a)
$$

and

$$
\varphi(\pi_A(N_A(a))) = \varphi(a).
$$

By Ivanov's theorem, λB is projective. Therefore, $\lambda\varphi$ is invertible. Let $\delta : \lambda B \to \lambda A$ denote an inverse (homomorphism).

$$
\begin{array}{ccc}
A & \xrightarrow{\ \varphi\ } & B \\[2pt]
{\scriptstyle \pi_A}\Big\uparrow & {\scriptstyle \pi_B}\Big\uparrow\Big\downarrow{\scriptstyle N_B} & \\[2pt]
\lambda A & \underset{\delta}{\overset{\lambda\varphi}{\rightleftarrows}} & \lambda B
\end{array}
$$

Using $\pi_B \circ \lambda\varphi = \varphi \circ \pi_A$, it is routine to check that $\varphi \circ [\pi_A \circ \delta \circ N_B] = id_B$. As all factors are monotone and preserve disjointness, $\varepsilon = \pi_A \circ \delta \circ N_B$ is the sought-for inverse. $\square$

The following result will be needed in section 5.5

Proposition 3.2.8 *The functor λ reflects and strongly preserves regular embeddings.*

Proof. We use the regularity test provided by 1.2.16(5). To prove REFLECTION we assume $\lambda A \leq_{reg} \lambda B$ and establish $A \leq_{reg} B$. Let $0 < b \in B$ be given. Then $\emptyset \neq N(b) \in \lambda B$ and there is a non-zero element of λA, which, by 3.2.2, we can assume to be of the form $N(a_1, \ldots, a_n)$, such that

(*) $\quad \lambda A \restriction (N(a_1, \ldots, a_n) \setminus N(b)) = \{\emptyset\}$

As $N(\vec{a}) \neq \emptyset$, all a_i are non-zero. Therefore, we will be done if we have proved

CLAIM 1. $A \restriction (a_i - b) = \{0\}$, *for some* i.

If that were false, we could pick $0 < c_i \in A$ such that $c_i \leq a_i - b$. Putting $c = c_1 \vee \ldots \vee c_n$, we would get from 3.2.3

$$\emptyset \neq N(c, a_1, \ldots, a_n) \subseteq N(a_1, \ldots a_n) \text{ and} N(c, a_1, \ldots, a_n) \cap N(b) = \emptyset.$$

As $N(c, a_1, \ldots, a_n) \in \lambda A$, these relations contradict (*).

Next we prove PRESERVATION. Assuming $A \leq_{reg} B$ we show that $\lambda A \leq_{reg} \lambda B$. Let $\emptyset \neq N(b_1, \ldots, b_n) \in \lambda B$ be given. We find a non-zero $N(a_1, \ldots, a_n) \in \lambda A$ such that $\lambda A \restriction (N(\vec{a}) \setminus N(\vec{b})) = \{\emptyset\}$. That will be sufficient, by 3.2.2 and 1.2.16(6).

For each $I \subseteq \{1, \ldots, n\}$ we put $b_I = \bigwedge_{i \in I} b_i$ and pick, using $A \leq_{reg} B$, elements $a_I \in A$ such that

(**) $\quad A \restriction (a_I - b_I) = \{0\}$.

It is clearly understood that $a_I > 0$ whenever $b_I > 0$. The following two claims show that if we put $a_i = \bigvee \{a_I : i \in I\}$, then $N(a_1, \ldots, a_n)$ is as desired.

CLAIM 2. $N(a_1, \ldots, a_n) \neq \emptyset$

By 3.2.3, this is equivalent to $a_i \wedge a_j \neq 0$ for all i, j. From $N(\vec{b}) \neq \emptyset$ we know $0 \neq b_i \wedge b_j = b_{\{i,j\}}$, hence $0 \neq a_{\{i,j\}} \leq a_i \wedge a_j$.

CLAIM 3. $\lambda A \restriction (N(\vec{a}) \setminus N(\vec{b})) = \{\emptyset\}$

Indeed, considering $\vec{c} \in A$ such that $N(\vec{c}) \cap N(\vec{b}) = \emptyset$, we have to show that $N(\vec{c}) \cap N(\vec{a}) = \emptyset$. By 3.2.3, $c_j \wedge b_i = 0$ for some $1 \leq i, j \leq n$. Then $c_j \wedge b_I = 0$ whenever $i \in I$. From (**) it follows that $c_j \wedge a_I = 0$ whenever $i \in I$, hence $c_j \wedge a_i = 0$, which yields $N(\vec{c}) \cap N(\vec{a}) = \emptyset$, as desired.

Preservation is proved and we go over to STRONG PRESERVATION. By induction on n we prove that if $A_1, \ldots, A_n$ are regular subalgebras of B, then for each $\emptyset \neq N(b_1, \ldots, b_k) \in \lambda B$ there are elements $a_1, \ldots, a_m \in A_1 \cup \ldots \cup A_n$ such that

(+) $\quad N(\vec{a}) \neq \emptyset \quad$ and $\quad \langle \lambda A_1 \cup \ldots \cup \lambda A_n \rangle \restriction (N(\vec{a}) \setminus N(\vec{b})) = \{\emptyset\}$.

For $n = 1$ this has just been proved. Assume that the assertion is true for n and consider $n + 1$ regular subalgebras $A_1, \ldots, A_n, A_{n+1}$ of B. Let $\emptyset \neq N(\vec{b}) \in \lambda B$ be given and choose $\vec{a} \in A_1 \cup \ldots \cup A_n$ according to the induction hypothesis, i.e. such that (+) holds. Then $N(a_1, \ldots, a_m, b_1, \ldots b_k) = N(\vec{a}) \cap N(\vec{b}) \neq \emptyset$ and we can apply the induction hypothesis (in fact the case $n = 1$) again to choose $e_1, \ldots, e_p \in A_{n+1}$ such that

$$(++) \quad N(\vec{e}) \neq \emptyset \quad \text{and} \quad \lambda A_{n+1} \restriction (N(\vec{e}) \setminus N(\vec{a}, \vec{b})) = \{\emptyset\}.$$

Notice that $N(\vec{e}, \vec{a}) = N(\vec{e}) \cap N(\vec{a}) \neq \emptyset$ because even $N(\vec{e}) \cap N(\vec{a}, \vec{b}) \neq \emptyset$. Therefore, we shall be ready if we establish

CLAIM 4. $(\lambda A_1 \cup \ldots \cup \lambda A_n \cup \lambda A_{n+1}) \restriction (N(\vec{e}, \vec{a}) \setminus N(\vec{b})) = \{\emptyset\}$

By 3.2.2, it is sufficient to consider any $N(\vec{c}, \vec{d}) \in (\lambda A_1 \cup \ldots \cup \lambda A_n \cup \lambda A_{n+1})$ such that $N(\vec{c}, \vec{d}) \cap N(\vec{b}) = \emptyset$ and to show that

$$\emptyset = N(\vec{c}, \vec{d}) \cap N(\vec{e}, \vec{a}) = N(\vec{c}) \cap N(\vec{e}) \cap N(\vec{d}) \cap N(\vec{a}).$$

Here it is understood, that $\vec{c} \in A_{n+1}$ and $\vec{d} \in A_1 \cup \ldots \cup A_n$. By 3.2.3, $N(\vec{c}, \vec{d}) \cap N(\vec{b}) = \emptyset$ implies that eiter $c_i \wedge b_j = 0$ or $d_i \wedge b_j = 0$ for suitable i, j. In the first case we get $N(\vec{c}) \cap N(\vec{a}, \vec{b}) = \emptyset$, hence $N(\vec{c}) \cap N(\vec{e}) = \emptyset$, by (++). In the second case, $N(\vec{d}) \cap N(\vec{a}) = \emptyset$, by (+). The claim is proved and, thereby, the proposition. $\square$

DIGRESSION

In connection with Ivanov's theorem we want to mention another functor. Topologists start from a space X and define γX to be the subspace of $exp\,exp\,X$ consisting of those closed collections $\mathcal{F}$ of closed subsets of X that satisfy

$$F \subseteq G \ \& \ F \in \mathcal{F} \implies F \in \mathcal{F}.$$

In the Boolean algebra setting things are easier. Let A be a Boolean algebra and denote by $G(A)$ the subspace of 2^A consisting of all monotone mappings that preserve 0 and 1. γA is then defined to be $Clop(GA)$. The rest of the definitions and proofs is in analogy to what we did for λ. The following theorem was first proved by Moisseev.

Information 3.2.9 ([41], Theorem 3) A is rc-filtered iff γA is projective.

Notice that $G(A)$ is a sublattice of $(2; \wedge, \vee)^A$. One can prove that $Ult\,A$ is a retract of $G(A)$ (in the topological sense) provided that $Ult\,A$ admits a distributive lattice structure (which is necessarily profinite). Hence

Information 3.2.10 If A is rc-filtered and $Ult\,A$ admits a distributive lattice structure, then A is projective.

It seems to be unknown whether all Stone spaces of projective Boolean algebras (i.e. zero-dimensional Dugundji spaces) admit distributive lattice structures. In [25] it is proved that they admit (not necessarily distributive) profinite lattice structures.

3.3   Exponentials

Our main aim in this section is to finally give an example of an rc-filtered Boolean algebra that is not projective.

The exponential or hyperspace of a topological space X is the set of all non-empty closed subsets of X equipped with the Vietoris topology. Below we define a purely algebraic concept[3], which under Stone duality corresponds to the topological exponential. The interested reader may consult [24] for more details and the proof that the topological and algebraic concepts are dual.

In this section we consider Boolean algebras as linear algebras over the field $\mathbf{F}_2$ with two elements (cf. 8).

Definition 3.3.1 Let A be given. We first forget about the addition in A reducing it to a semigroup with 0 and 1. Then we form the semigroup algebra with coefficients in the field $\mathbf{F}_2$. The result is a Boolean algebra which will be denoted by $exp\,A$ and called the *exponential* of A. □

The elements of $exp\,A$ are all finite formal sums $a_1 \oplus \ldots \oplus a_n$[4] of pairwise distinct non-zero elements of A together with 0 (to be identified with $0 \in A$)[5]. Equality, addition and multiplication are defined in the obvious way. Notice that the non-zero elements of A constitute a base of the vector space $exp\,A$ and that $a \oplus a = 0$ for all $a \in A$.

The reader should verify the following *universality property* that could be used to define exp in the style of category theory.

Lemma 3.3.2 *If $\varphi : A \to B$ is multiplicative, i.e. preserves $\cdot, 0, 1$, then the formula*

$$a_1 \oplus \ldots \oplus a_n \;\mapsto\; \varphi(a_1) + \ldots + \varphi(a_n)$$

defines a Boolean algebra homomorphism $\overline{\varphi} : exp\,A \to B$.

On homomorphisms exp acts the obvious way.

Definition 3.3.1 (continued) If $\varphi : A \to B$ is a Boolean algebra homomorphism, then $exp\,\varphi : exp\,A \to exp\,B$ is defined by

$$exp\,\varphi(a_1 \oplus \ldots \oplus a_n) = \varphi(a_1) \oplus \ldots \oplus \varphi(a_n). \quad □$$

It is a matter of routine to check that all our definitions are correct and exp becomes a functor. It is equally easy to see that exp preserves cardinalities. If $A \leq B$, then $exp\,A$ is obviously isomorphic to the subalgebra of $exp\,B$ consisting of those formal sums whose terms are in A. As each element of $exp\,A$ involves only finitely many elements of A, the functor exp is continuous. It will cost some work to check that exp preserves relative completeness of embeddings.

---

[3] Yet another approach to exponentials is sketched at the end of this section.

[4] As the operations of A and $exp\,A$ are considered at the same time, we use $\oplus, \odot$ to distinguish the latter. $\preceq$ denotes the Boolean partial order in $exp\,A$.

[5] Not to bother about the, usually trivial, case 0, we subsume it under the form $a_1 \oplus \ldots \oplus a_n$, for $n = 0$.

We have to introduce some notation. For $F \subseteq A$ we put

$$I(F) = \{a_1 \oplus \ldots \oplus a_n : \{i : a_i \in F\} \text{ has an even number of elements}\}$$

and

$$U(F) = \{a_1 \oplus \ldots \oplus a_n : \{i : a_i \in F\} \text{ has an odd number of elements}\}.$$

Lemma 3.3.2 *Let F be a proper (i.e. $0 \notin F$) filter of A.*

(1) *$I(F)$ is a prime ideal of $\exp A$.*

(2) *Each prime ideal of $\exp A$ can be written in the form $I(F)$ for a suitable filter F of A.*

(3) *$U(F)$ is an ultrafilter of $\exp A$.*

(4) *Each ultrafilter of $\exp A$ can be written in the form $U(F)$ for a suitable filter F of A.*

Proof. (1) Define a mapping $\varphi : A \to 2$ by setting $\varphi(a) = 1$ iff $a \in F$. It is easily checked that φ is multiplicative. By the universality property 3.3.2, φ extends to a homomorphism $\overline{\varphi} : \exp A \to 2$. But $I(F)$ is defined to be the kernel of this extension.

(2) Let J be a prime ideal of $\exp A$. Put $F = \{a \in A : 1 \oplus a \in J\}$. Let us check that F is a filter of A. If $a, b \in F$, i.e. $1 \oplus a, 1 \oplus b \in J$, then $(1 \oplus a) \oplus (1 \oplus b) \oplus (1 \oplus a) \odot (1 \oplus b) = 1 \oplus ab \in J$, i.e. $ab \in F$. If $a \leq b$ and $a \in F$, then $1 \oplus a \in J$ and $1 \oplus b = (1 \oplus a) \odot (1 \oplus b) \in J$, hence $b \in F$.

As both ideals are prime, it is sufficient to verify $I(F) \subseteq J$. Take any $a_1 \oplus \ldots \oplus a_{2k} \oplus a_{2k+1} \oplus \ldots \oplus a_n \in I(F)$, where $a_1, \ldots, a_{2k} \in F$ and $a_{2k+1}, \ldots, a_n \notin F$. Then $1 \oplus a_1, \ldots, 1 \oplus a_{2k} \in J$ and, as J is prime, $a_{2k+1}, \ldots a_n \in J$. As an even number of 1's gets cancelled, it follows that

$$a_1 \oplus \ldots \oplus a_{2k} \oplus a_{2k+1} \oplus \ldots a_n = (1 \oplus a_1) \oplus \ldots \oplus (1 \oplus a_{2k}) \oplus a_{2k+1} \oplus \ldots a_n \in J.$$

Noticing that $1 \in A$ acts as unit of $\exp A$ so that $1 \oplus x$ is the complement of $x \in \exp A$, assertions (3) and (4) follow from (1) and (2). $\square$

More notation is needed. For $a_1, \ldots, a_n \in A$ we put

$$v(a_1, \ldots, a_n) = (a_1 \vee \ldots \vee a_n) \odot (1 \oplus -a_1) \odot \ldots \odot (1 \oplus -a_n)$$

and denote by $V(A)$ the set

$$\{v(a_1, \ldots, a_n) : a_1, \ldots, a_n \in A \text{ are non-zero and pairwise disjoint}\}.$$

It is obvious that $v(a_1, \ldots, a_n)$ does not depend on the order of its arguments. It is equally clear that $v(a_1, \ldots, a_n)$ becomes 0 if one a_i is 0. If $n = 1$, then $v(a) = a \odot (1 \oplus -a) = a \oplus (a \odot -a) = a \oplus (a \wedge -a) = a \oplus 0$ is simply a. It is sometimes reasonable to write $v(a)$ instead of a to stress that it is considered as an element of $\exp A$.

Lemma 3.3.3 *Assume that* $a_1, \ldots, a_n, b_1, \ldots b_m \in A$ *are non-zero.*

(1) $v(a_1, \ldots, a_n) \neq 0$

(2) *If* $n = m$ *and* $a_i \leq b_i$ *for all* i, *then* $v(a_1, \ldots, a_n) \preceq v(b_1, \ldots, b_n)$

(3) $v(a_1, \ldots, a_n) \odot v(b_1, \ldots, b_m) = 0$ *iff either one* a_i *is disjoint from all* b_j *or one* b_j *is disjoint from all* a_i.

Proof. (1) Let F denote the (proper) filter of A generated by $a_1 \vee \ldots \vee a_n$. Notice that all factors of the product making up $v(a_1, \ldots, a_n)$ belong to $U(F)$. So, $v(a_1, \ldots, a_n) \in U(F)$, which, as an ultrafilter, does not contain 0.

(2) From the assumption we get immediately that

$$(a_1 \vee \ldots \vee a_n) \odot (b_1 \vee \ldots \vee b_n) = (a_1 \vee \ldots \vee a_n) \wedge (b_1 \vee \ldots \vee b_n) = (a_1 \vee \ldots \vee a_n),$$

i.e. $(a_1 \vee \ldots \vee a_n) \preceq (b_1 \vee \ldots \vee b_n)$,   and, for all $1 \leq i \leq n$,

$$(1 \oplus -a_i) \odot (1 \oplus -b_i) = 1 \oplus -a_i \oplus -b_i \oplus -(a_i \vee b_i) = 1 \oplus -a_i \oplus -b_i \oplus -b_i$$

$= 1 \oplus -a_i$, i.e. $(1 \oplus -a_i) \preceq (1 \oplus -b_i)$. It remains to multiply the obtained inequalities.

(3) Substituting the definition of v, we get $v(a_1, \ldots, a_n) \odot v(b_1, \ldots, b_m) =$

$$(a_1 \vee \ldots \vee a_n) \odot (b_1 \vee \ldots \vee b_m) \odot$$
$$(1 \oplus -a_1) \odot \ldots \odot (1 \oplus -a_n) \odot (1 \oplus -b_1) \odot \ldots \odot (1 \oplus -b_m).$$

Call this expression the long product. If a_i, say, is disjoint from all b_j, then $(1 \oplus -a_i) \odot (b_1 \vee \ldots \vee b_m) = 0$ and the long product becomes zero. For the other direction we assume that the long product is zero and consider the filter F of A generated by $(a_1 \vee \ldots \vee a_n) \wedge (b_1 \vee \ldots \vee b_m)$. If that intersection is empty, we are done already. Otherwise, F is a proper filter and, by 3.3.2, $U(F)$ an ultrafilter of $exp\,A$. As the long product does not belong to $U(F)$, one of the factors does not belong to $U(F)$. As the first two factors are in $U(F)$, this can only be one of $(1 \oplus -a_i)$ or $(1 \oplus -b_j)$, i.e. either $-a_i \in F$ or $-b_j \in F$, which easily implies the assertion. $\square$

Lemma 3.3.4 *Each non-zero element of* $exp\,A$ *is the union of finitely many elements in* $V(A)$.

Proof. As each element of $exp\,A$ involves only finitely many elements of A, we can assume that A itself is finite. Let $a_1, \ldots, a_n$ be the atoms of A. An easy calculation shows that $exp\,A$ has exactly 2^{2^n-1} elements, hence $2^n - 1$ atoms. On the other hand, there are $2^n - 1$ formally different elements of the form $v(a_{i_1}, \ldots, a_{i_k})$. By 3.3.3 they are non-zero and pairwise disjoint. So, they are the atoms of $exp\,A$. $\square$

Proposition 3.3.5 *The functor* exp *preserves and reflects relative completeness of embeddings, i.e.* $A \leq_{rc} B$ *iff* $exp\,A \leq_{rc} exp\,B$.

Proof. We start with PRESERVATION. Assume $A \leq_{rc} B$. By 1.2.3 and 3.3.4 it is sufficient to find a minimal element of $exp\,A$ above $v(b_1, \ldots, b_n)$, where $b_1, \ldots, b_n$ are pairwise disjoint non-zero elements of B. The obvious idea works. Let a_i be minimal in A above b_i. Then $v(b_1, \ldots, b_n) \preceq v(a_1, \ldots, a_n)$, by 3.3.3(2). To establish minimality, we consider an arbitrary element $v(c_1, \ldots, c_m) \in V(A)$ which intersects $v(a_1, \ldots a_n)$ and prove that it also intersects $v(b_1, \ldots b_n)$.

Assume, on the contrary, that $v(c_1, \ldots c_m) \odot v(b_1, \ldots b_n) = 0$. By 3.3.3(3), either c_1, say, is disjoint from all b_j, or b_1, say, is disjoint from all c_i. In the first case, c_1 is disjoint from all a_j, by minimality of a_j above b_j. In the second case, a_1 is disjoint from all c_i, also by minimality. So in both cases, $v(c_1, \ldots c_m) \odot v(a_1, \ldots a_n) = 0$, contradiction.

Now we prove REFLECTION. Assume $A \leq B$ and $exp\,A \leq_{rc} exp\,B$. Let $b \in B$ be given and take the maximal element, w say, of $exp\,A$ below $v(b)$. By 3.3.4, it can be written as $w = \bigvee_{i=1}^{n} v(a_1^i, \ldots, a_{m_i}^i)$, where all a_j^i are non-zero. We put $a = \bigvee_{i=1}^{n} (a_1^i \vee \ldots \vee a_{m_i}^i)$ and prove that a is maximal in A below b.

CLAIM 1. $a \leq b$

As the numeration plays no role, it is sufficient to establish $a_1^1 \leq b$. Consider the element $v(a_1^1 - b, a_2^1, \ldots, a_{m_1}^1)$. By 3.3.3(2), it is below $v(a_1^1, a_2^1, \ldots, a_{m_1}^1) \preceq v(b)$ and, by 3.3.3(3) it is disjoint from $v(b)$. Consequently, $v(a_1^1 - b, a_2^1, \ldots, a_{m_1}^1) = 0$. As all a_j^i are non-zero, we must have $a_1^1 - b = 0$, by 3.3.3(1).

CLAIM 2. a is maximal in A below b.

Indeed, consider any $c \in A$ such that $c \leq b$. Then $v(c - a) \in exp\,A$ and $v(c - a) \preceq v(c) \preceq v(b)$. It follows that $v(c - a) \preceq w$. On the other hand, by 3.3.3(3), $v(c - a) \odot v(a_1^i, \ldots, a_{m_i}^i) = 0$, for all i. Hence, $v(c - a) \odot w = 0$. It follows that, $v(c - a) = 0$, hence $c \leq a$. $\square$

From 3.1.1 and the properties of exp that we established so far we immediately obtain

Corollary 3.3.6 *A is rc-filtered iff $exp\,A$ is rc-filtered.*

Our next aim is to study exponentials of free algebras. From 3.3.6 it follows that they are rc-filtered. Some more work is needed before we can say more.

Let us consider the following situation. $A \leq B$, F is a proper filter of A, b^* is an element of B that is independent of A, i.e. $a \wedge b^*$ and $a - b^*$ are non-zero for all non-zero $a \in A$.[6] By $\hat{I}(F)$ we denote the ideal of $exp\,B$ generated by the prime ideal $I(F)$ of $exp\,A$. In the following lemma we write $v(x)$ instead of x to stress that we are in exp.

Lemma 3.3.7 *In the above situation the following assertions hold.*

---

[6] Notice that for $a, c \in A$, $b^* \wedge a \leq c$ or $-b^* \wedge a \leq c$ imply $a \leq c$.

(1) $v(b) \in \hat{I}(F)$ iff there is some $a \in A \setminus F$ such that $b \leq a$.

(2) $\hat{I}(F)$ is not prime.

(3) $v(b^* + a) \notin \hat{I}(F)$, for all $a \in A$.

(4) If a_1/F and a_2/F are non-zero disjoint elements of A/F,
then $v(b^* + a_1) \odot v(b^* + a_2) \in \hat{I}(F)$.

(5) $c(A/F) \leq c(exp\, B/\hat{I}(F))$, where c denotes cellularity.

Proof. (1) If $b \leq a \in A \setminus F$, then, by definition, $v(a) \in I(F)$. As $v(b) \preceq v(a)$,
we get $v(b) \in \hat{I}(F)$.

Assume $v(b) \in \hat{I}(F)$, then

$$v(b) \preceq v(a_1) \oplus \ldots \oplus v(a_{2k}) \oplus v(c_1) \ldots \oplus v(c_m),$$

where the a's belong to F and the c's don't. Rewriting this inequality, we get
$v(b) = v(a_1 \wedge b) \oplus \ldots \oplus v(c_m \wedge b)$. By linear independence, either $v(b) = v(a_i \wedge b)$
or $v(b) = v(c_j \wedge b)$. In the latter case we have, $b \leq c_j \in A \setminus F$, as desired. In the
former case, we get, assuming $i = 1$,

$$0 = v(a_2 \wedge b) \oplus \ldots \oplus v(a_{2k} \wedge b) \oplus v(c_1 \wedge b) \oplus \ldots \oplus v(c_m \wedge b)$$

As an odd number of a's remained, they cannot cancel each other. So one
$v(a_i \wedge b)$ has to be equal to one $v(c_j \wedge b)$ (or zero, which is covered by allowing
$c_j = 0$). But, $a_i \wedge b = c_j \wedge b$ implies $b \leq -a_i \vee c_j$, which is not in F. For,
otherwise, $a_i \wedge (-a_i \vee c_j) \leq c_j$ were in F, which is not the case.

(2) As the independent element b^* is not covered by any element of A, except
$1 \in F$, it follows from (1) that $v(b^*) \notin \hat{I}(F)$. But $-b^*$ is as independent as b^*.
So $v(-b^*) \notin \hat{I}(F)$, too. If $\hat{I}(F)$ were prime then one of the disjoint elements
$v(b^*), v(-b^*)$ had to be in $\hat{I}(F)$.

(3) Assume, by contradiction, $v(b^* + a) \in \hat{I}(F)$. Then, by (1), $b^* + a \leq d$ for some
$d \in A \setminus F$. It follows that $b^* \wedge -a \leq d$ and $-b^* \wedge a \leq d$, which, by independence,
imply, $a \leq d$ and $-a \leq d$. Consequently, $d = 1 \in F$, contradiction.

(4) Let $f \in F$ be such that $a_1 \wedge a_2 \wedge f = 0$. By definition,

$$v(b^* + a_1) \odot v(b^* + a_2) = v((b^* + a_1) \cdot (b^* + a_2)).$$

An elementary calculation shows

$$(b^* + a_1) \cdot (b^* + a_2) \cdot (a_1 + a_2) = 0,$$

which yields

$$(b^* + a_1) \cdot (b^* + a_2) \leq 1 + a_1 + a_2.$$

By (1), it will be sufficient to check that $1 + a_1 + a_2$ does not belong to F. If
it would, then so would $g = f \cdot (1 + a_1 + a_2)$. But $a_1 \cdot g = 0$, so we would get
$a_1/F = 0/F$, contradiction.

(5) follows immediately from (3) and (4). $\square$

Corollary 3.3.8 *For infinite X, all ultrafilters of $exp \, \mathrm{Fr} \, X$ have character $|X|$.*

Proof. Assume the contrary and take an ultrafilter U of $exp \, \mathrm{Fr} \, X$ which is generated by its intersection with a subalgebra $A \leq exp \, \mathrm{Fr} \, X$ of small power. Using the continuity of exp, there is no problem to find some $Y \subseteq X$ also of small power such that $A \leq exp \, \mathrm{Fr} \, Y \leq exp \, \mathrm{Fr} \, X$.[7] Clearly, $U \cap exp \, \mathrm{Fr} \, Y$ is an ultrafilter of $exp \, \mathrm{Fr} \, Y$ and generates U. On the other hand, $\mathrm{Fr} \, X$ contains elements that are independent of $\mathrm{Fr} \, Y$. So, no ultrafilter of $exp \, \mathrm{Fr} \, Y$ generates an ultrafilter of $exp \, \mathrm{Fr} \, X$, by (2) of the above lemma and 3.3.2(4). $\square$

We know now that $exp \, \mathrm{Fr} \, \omega$ and $exp \, \mathrm{Fr} \, \omega_1$ are rc-filtered, hence projective (because of their power $\leq \aleph_1$) and have only ultrafilters of full character. Consequently, by Ščepins theorem 1.4.11, $exp \, \mathrm{Fr} \, \omega \simeq \mathrm{Fr} \, \omega$ and $exp \, \mathrm{Fr} \, \omega_1 \simeq \mathrm{Fr} \, \omega_1$. The latter result has first been proved by Sirota [57]. The situation changes if cardinalities become bigger.

Proposition 3.3.9 *If $|X| > \aleph_1$, then $exp \, \mathrm{Fr} \, X$ is not projective.*

Proof. Assume the contrary. Then, as for ω and ω_1 above, there would be an isomorphism $\varphi : exp \, \mathrm{Fr} \, X \longrightarrow \mathrm{Fr} \, X$. Starting with an arbitrary $Y_0 \subseteq X$ of power $\aleph_1$ we can inductively choose subsets Y_n of X such that

(1) $|Y_n| = \aleph_1$,

(2) $Y_n \subseteq Y_{n+1}$,

(3) $\mathrm{Fr} \, Y_n \subseteq \varphi(exp \, \mathrm{Fr} \, Y_{n+1})$, and

(4) $\varphi(exp \, \mathrm{Fr} \, Y_n) \subseteq \mathrm{Fr} \, Y_{n+1}$.

This is possible because of the continuity of exp and Fr. For $Y = \bigcup_{n<\omega} Y_n$ it holds that $|Y| = \aleph_1$ and $\varphi(exp \, \mathrm{Fr} \, Y) = \mathrm{Fr} \, Y$.

It follows that the embeddings $exp \, \mathrm{Fr} \, Y \leq exp \, \mathrm{Fr} \, X$ and $\mathrm{Fr} \, Y \leq \mathrm{Fr} \, X$ are isomorphic. To see that this is impossible, we take a filter F of $\mathrm{Fr} \, Y$ such that $\mathrm{Fr} \, Y / F$ does not have the ccc. Then $I(F)$ is a prime ideal of $exp \, \mathrm{Fr} \, Y$ and $(exp \, \mathrm{Fr} \, X)/\hat{I}(F)$ does not have the ccc either, by 3.3.7(5). On the other hand, if I is a prime ideal of $\mathrm{Fr} \, Y$, then $(\mathrm{Fr} \, X)/\hat{I} \simeq \mathrm{Fr} \, (X \setminus Y)$ is free and has the ccc. $\square$.

The above result now enables us to say exactly when exponentials are projective.

Corollary 3.3.10 *$exp \, A$ is projective iff A is projective and $|A| \leq \aleph_1$.*

Proof. Assume first that $exp \, A$ is projective. Applying the universality property 3.3.2 to the identical mapping $id : A \to A$ yields a, necessarily surjective, homomorphism $\overline{id} : exp \, A \to A$. Clearly, the mapping sending $a \in A$ to $v(a) \in exp \, A$ is a right-inverse of $\overline{id}$. Moreover, v preserves zero and intersections, which yields the projectivity of A, by 2.4.3.

---

[7] We identify $\mathrm{Fr} \, Y$ with the subalgebra (Y) of $\mathrm{Fr} \, X$.

To see that $|A| \leq \aleph_1$, we derive a contradiction from the assumption $|A| \geq \aleph_2$. If this were true, 2.7.9 would imply that $\mathrm{Fr}\,\omega_2$ is a retract of A. As functors preserve retractions, we would get the non-projective algebra $exp\mathrm{Fr}\,\omega_2$ as a retract of the projective $exp\,A$. This contradicts 1.5.6(2).

The other direction is easier. If A is projective of cardinality $\leq \aleph_1$, then $exp\,A$ is rc-filtered, by 3.3.6, and also of cardinality $\leq \aleph_1$. By 2.2.7, the two classes coincide for small algebras, hence $exp\,A$ is even projective. $\square$

Based on the same idea but with a more complicated technique as used in the proof of 3.3.9, we now establish the following more general result.

Theorem 3.3.11 ([47]) *If $|X| > \aleph_1$, then $exp\,\mathrm{Fr}\,X$ is not isomorphic to a subalgebra of a free Boolean algebra.*

We separate part of the argument as the following somewhat lengthy and somewhat abstract

Lemma 3.3.12 *Assume that E, B, E_1, B_1 are situated like this:*

$$\begin{array}{ccc} E & \leq & B \\ \mathrm{VI} & & \mathrm{VI} \\ E_1 & \leq & B_1 \end{array}$$

Assume further that I is an ideal of B such that

(1) $E \cap I = \{0\}$, *and*

(2) *for each $b \in B_1$ the ideal*

$$E{\restriction}(b \vee I) = \{e \in E : e \leq b \vee i \text{ for some } i \in I\}$$

is generated by its intersection with E_1.

Then for each prime ideal P of E_1 there exists a prime ideal Q of B_1 such that $\hat{P} = \hat{Q} \cap E$, where $\hat{P}$ and $\hat{Q}$ denote the ideals generated by P and Q in E and B, respectively. In particular, $E/\hat{P}$ embeds into $B/\hat{Q}$.

Proof. For the sake of this proof we call an ideal Q of B_1 *clean* if $Q \cup I$ generates a proper ideal of B, i.e. if there are no $q \in Q$ and $i \in I$ such that $q \vee i = 1$.

Notice that the ideal $\bar{P}$ generated by P in B_1 is clean. Indeed, otherwise, $p \vee i = 1$ for some $p \in P, i \in I$. But then, $-p \leq i$, hence $-p \in I \cap E = \{0\}$ and $p = 1$, contradiction.

By Zorn's Lemma, we can pick an ideal Q that is maximal with respect to the properties of being clean and containing P. We show that Q is as desired.

CLAIM 1. *Q is a prime ideal of B_1.*

Indeed, let $b \in B_1$ be given. We have to show that either b or $-b$ belongs to Q. If neither does, then, by maximality,

$$b \vee q \vee i = 1 \quad \text{and} \quad -b \vee r \vee j = 1$$

for suitable $q, r \in Q$ and $i, j \in I$. It follows that

$$(q \vee r) \vee (i \vee j) = 1.$$

As $q \vee r \in Q$ and $i \vee j \in I$, this contradicts the cleanness of Q.

CLAIM 2. $\hat{P} = \hat{Q} \cap E$.

As $\hat{P} \subseteq \hat{Q} \cap E$ follows trivially from $P \subseteq Q$, it remains to establish the reverse inclusion.

Take any $e \in \hat{Q} \cap E$. Then $e \leq q$ for some $q \in Q \subseteq B_1$. Condition (2) yields $e' \in E_1$ and $i \in I$ such that $e \leq e' \leq q \vee i$, hence $-e' \vee q \vee i = 1$. It is sufficient to show $e' \in P$. But, if that were not the case, then $-e' \in P \subseteq Q$, hence $-e' \vee q \in Q$, which, together with $-e' \vee q \vee i = 1$, contradicts the cleanness of Q. $\square$

Now we **prove the theorem**. It will be sufficient to derive a contradiction from the assumption that $\varphi : exp \operatorname{Fr} \omega_2 \longrightarrow \operatorname{Fr} \omega_2$ is an injective homomorphism. Let Y be a subset of ω_2. Let us denote the subalgebra $\varphi(exp \operatorname{Fr} Y)$ of $\operatorname{Fr} \omega_2$ by $E(Y)$.

Being isomorphic to $exp \operatorname{Fr} \omega_2$, $E(\omega_2)$ is rc-filtered, by 3.3.6. Therefore (2.2.7), it has the Bockstein separation property. Zorn's Lemma allows us to pick an ideal I of $\operatorname{Fr} \omega_2$ which is maximal with respect to the property $E(\omega_2) \cap I = \{0\}$. By 1.2.21 and the (BSP), the ideal $E(\omega_2) \restriction (b \vee I)$ is countably generated, for all $b \in \operatorname{Fr} \omega_2$.

It is then routine (cf. the proof of the proposition above) to construct a subset $Y \subseteq \omega_2$ such that the following conditions are satisfied.

(1) $|Y| = \aleph_1$,

(2) $E(Y) \subseteq \operatorname{Fr} Y$, and

(3) for each $b \in \operatorname{Fr} Y$, the ideal $E(\omega_2) \restriction (b \vee I)$ is generated by its intersection with $E(Y)$.

So, with I and the algebras

$$
\begin{array}{ccc}
E(\omega_2) & \leq & \operatorname{Fr} \omega_2 \\
\mathrm{VI} & & \mathrm{VI} \\
E(Y) & \leq & \operatorname{Fr} Y
\end{array}
$$

we are in the situation of the lemma. It follows that for each prime ideal P of $E(Y)$ there is a prime ideal Q of $\operatorname{Fr} Y$ such that $E(\omega_2)/\hat{P}$ embeds into $\operatorname{Fr} \omega_2/\hat{Q}$. The latter quotient is always isomorphic to $\operatorname{Fr} \omega_2$ and has, therefore, the ccc. It follows that $E(\omega_2)/\hat{P}$ has the ccc, for all prime ideals P of $E(Y)$. But, we know from 3.3.7(5) that this is not the case. $\square$

The ideas developped in the previous proof were used to establish the following result.

Information 3.3.13 ([50]) *Let A be an infinite subalgebra of a free Boolean algebra. For $\exp A$ to have a homogeneous Stone space it is necessary and sufficient that A is isomorphic to either $\mathrm{Fr}\,\omega$ or $\mathrm{Fr}\,\omega_1$.*

In the rest of this section we give two results about the behaviour of exponentials with respect to regular embeddings. It should come as no surprise that

Proposition 3.3.14 *The functor $\exp$ preserves and reflects regular embeddings.*

The **proof** is very similar to that of 3.3.5 and left as an exercise. (Use the regularity test 1.2.16(5)).

The second, rather technical, result will be needed in the proof of 5.4.5 only. We put its proof here in the hope that the reader is more familiar now with the necessary calculations in exponentials.

Alongside the cellularity we know already, there is another cardinal function[8] involved in its formulation. For an embedding $C \leq D$ we put

$$\pi(D/C) = \min\{|X| : \langle C \cup X \rangle \leq_d D\}.$$

This number will be called the π-*weight* of D over C.

Proposition 3.3.15 (cf. Lemma 8 in [49]) *If $A \leq_{reg} B$, then*

$$\pi(\exp B \,/\, \exp A) \leq \pi(B/A) \cdot \sup\{c(A^{(n)}) : n < \omega\},$$

where $A^{(n)}$ denotes the free power $A \otimes A \otimes \ldots \otimes A$ with n factors.

Proof. For the sake of this proof we call a non-zero element of $A^{(n)}$ *distinguished* if it is of the form $d_1 \otimes d_2 \otimes \ldots \otimes d_n$ and for all $i \neq j$ either $d_i \wedge d_j = 0$ or $d_i = d_j$ is an atom of A. Distinguished elements will be useful because of the following two formulas

(1) $[\bigvee_{i=1}^{n}(e_i \wedge d_i)] \wedge [\bigvee_{i=1}^{n}(c_i \wedge d_i)] = \bigvee_{i=1}^{n}(e_i \wedge c_i \wedge d_i)$ and

(2) $[\bigvee_{i=1}^{n}(e_i \wedge c_i \wedge d_i)] - (e_j \wedge c_j) \leq -(c_j \wedge d_j)$

which hold if $d_1 \otimes d_2 \otimes \ldots \otimes d_n$ is distinguished and $e_1, \ldots, e_n \in A$, $c_1, \ldots, c_n \in B$ are such that $e_i \wedge d_i \neq 0 \neq c_i \wedge d_i$ for all i. Indeed, by distributivity,

$$[\bigvee_{i=1}^{n}(e_i \wedge d_i)] \wedge [\bigvee_{i=1}^{n}(c_i \wedge d_i)] = [\bigvee_{i=1}^{n}(e_i \wedge c_i \wedge d_i)] \vee [\bigvee_{i \neq j}(e_i \wedge d_i \wedge c_j \wedge d_j)].$$

---

[8] Its topological dual was introduced by Šapirovskii in 1976 under the name of 'index of reducibility'.

Therefore, to prove (1), we have to establish

$$e_i \wedge d_i \wedge c_j \wedge d_j \leq \bigvee_{i=1}^{n} (e_i \wedge c_i \wedge d_i)$$

for $i \neq j$. This is clear if the left-hand side is 0. Otherwise, $d_i \wedge d_j \neq 0$, hence, $d_i = d_j$ is an atom of A. As $e_i, e_j \in A$ and $e_i \wedge d_i$ and $e_j \wedge d_j$ are non-zero, we must have $e_i \wedge d_i = d_i = d_j = e_j \wedge d_j$, hence

$$e_i \wedge d_i \wedge c_j \wedge d_j = e_j \wedge d_j \wedge c_j \wedge d_j \leq \bigvee_{i=1}^{n} (e_i \wedge c_i \wedge d_i),$$

as desired. To prove (2) we first rewrite it as

$$\bigvee_{i=1}^{n} (e_i \wedge c_i \wedge d_i) \leq -(c_j \wedge d_j) \vee (e_j \wedge c_j) = -c_j \vee -d_j \vee (e_j \wedge c_j)$$

and denote the right-hand side by r. To see that the inequality holds we go through all the joinands of the left-hand side. Obviously, $e_j \wedge c_j \wedge d_j \leq e_j \wedge c_j \leq r$. If $i \neq j$ and $d_i \wedge d_j = 0$, then $e_i \wedge c_i \wedge d_i \leq -d_j \leq r$. If $i \neq j$ and $d_i \wedge d_j \neq 0$, then $d_i = d_j$ is an atom of A, hence, as before, $e_i \wedge d_i = d_i = d_j = e_j \wedge d_j$. Therefore,

$$e_i \wedge c_i \wedge d_i = e_j \wedge c_i \wedge d_j \leq e_j \leq -c_j \vee (e_j \wedge c_j) \leq r,$$

as desired.

Next we establish that there are many distinguished elements.

CLAIM 1. *The set of distinguished elements is dense in $A^{(n)}$.*

To see this, recall first that each non-zero element of $A^{(n)}$ can be written as a finite join of elements of the form $a_1 \otimes \ldots \otimes a_n$, with all a_i non-zero. So it is sufficient to construct a distinguished element below any given $a_1 \otimes \ldots \otimes a_n$. For each i let $M(i)$ be a maximal subset of $\{1, \ldots, n\}$ such that $i \in M(i)$ and $\bigwedge_{j \in M(i)} a_j \neq 0$. Put $a_i' = \bigwedge_{j \in M(i)} a_j$. As $a_i' \neq 0$ and $i \in M(i)$, we then have

$$0 \neq a_1' \otimes \ldots \otimes a_n' \leq a_1 \otimes \ldots \otimes a_n.$$

Moreover, for $i \neq j$ either $M(i) = M(j)$, then $a_i' = a_j'$, or $M(i) \neq M(j)$, then $a_i' \wedge a_j' = 0$, by maximality.

Next we choose non-zero elements $a_{i1}'', \ldots, a_{in}''$ below every a_i'. If a_i' contains atoms, we let $a_{i1}'' = \ldots = a_{in}''$ be one of them. If a_i' is atomless, we choose the a_{ik}'' pairwise disjoint. Moreover, we choose in such a way that $a_{i1}'' = a_{j1}'', \ldots, a_{in}'' = a_{jn}''$ whenever $a_i' = a_j'$. As the a_{ik}'' are non-zero and below a_i' we have

$$0 \neq a_{11}'' \otimes \ldots \otimes a_{nn}'' \leq a_1' \otimes \ldots \otimes a_n' \leq a_1 \otimes \ldots \otimes a_n.$$

Moreover, from the construction it is clear that $a_{11}'' \otimes \ldots \otimes a_{nn}''$ is distinguished. The claim is proved.

Now we fix for each $1 \leq n < \omega$ a maximal set of pairwise disjoint distinguished elements and call it $D^{(n)}$. From claim 1 we immediately get

CLAIM 2. *Every non-zero element of $A^{(n)}$ intersects some element of $D^{(n)}$.*

Next we fix a subalgebra C of B such that $|C| = \pi(B/A)$ and $\langle A \cup C \rangle \leq_d B$.

CLAIM 3. *The set $S =$*

$$\{v(a_1 \wedge c_1, \ldots, a_n \wedge c_n) : n < \omega,\ a_1, \ldots, a_n \in A,\ c_1, \ldots, c_n \in C,\ a_i \wedge c_i \neq 0\}$$

   is dense in $exp\, B$.

Indeed, by 3.3.4, the set of all $v(b_1 \ldots, b_n)$ with non-zero $b_i \in B$ is dense in $exp\, B$. From $\langle A \cup C \rangle \leq_d B$ it follows that below each b_i there is some non-zero $a_i \wedge c_i$, with $a_i \in A$ and $c_i \in C$. From 3.3.3 it follows that

$$0 \neq v(a_1 \wedge c_1, \ldots, a_n \wedge c_n) \preceq v(b_1, \ldots, b_n),$$

which proves the claim.

Now we introduce two subsets of $exp\, B$:

$X = \{v(a_1, \ldots, a_n) : n < \omega,\ a_1, \ldots, a_n \in A\}$ and
$Y = \{v(c_1 \wedge d_1, \ldots, c_n \wedge d_n) : n < \omega,\ c_1, \ldots, c_n \in C,\ d_1 \otimes \ldots \otimes d_n \in D^{(n)}\}$.

Then $|Y| \leq |C| \cdot \sup\{|D^{(n)}| : n < \omega\} \leq \pi(B/A) \cdot \sup\{c(A^{(n)}) : n < \omega\}$ and $X \subseteq exp\, A$. In view of claim 3, the proposition will be proved if we show that

CLAIM 4. *The set $\{x \odot y : x \in X,\ y \in Y\}$ is dense in S.*

Let $0 \neq s = v(a_1 \wedge c_1, \ldots a_n \wedge c_n) \in S$ be given. We find $x \in X$ and $y \in Y$ such that $0 \neq x \odot y \preceq s$. It is only here that regularity comes into play. It allows us to pick non-zero elements $e_1, \ldots, e_n \in A$ such that $A\!\restriction [e_i - (a_i \wedge c_i)] = \{0\}$ for all i. We must have $e_i \wedge a_i \neq 0$, for, otherwise, $0 \neq e_i \in A\!\restriction [e_i - (a_i \wedge c_i)] = \{0\}$. So, replacing e_i if necessary by $e_i \wedge a_i$, we shall assume

   (3) $e_i \leq a_i$, for all i.

By claim 2, $e_1 \otimes \ldots \otimes e_n$ intersects some $d_1 \otimes \ldots \otimes d_n \in D^{(n)}$, i.e.

   (4) $e_i \wedge d_i \neq 0$, for all i.

We now put $x = v(e_1 \wedge d_1, \ldots, e_n \wedge d_n)$ and $y = v(c_1 \wedge d_1, \ldots, c_n \wedge d_n)$ and show that these elements are as desired.

All meets $e_i \wedge d_i \wedge c_i$ are non-zero, for, otherwise, by (4)

$$0 \neq e_i \wedge d_i \in A\!\restriction [e_i - (a_i \wedge c_i)] = \{0\}.$$

So, by 3.3.3, $0 \neq v(e_1 \wedge d_1 \wedge c_1, \ldots, e_n \wedge d_n \wedge c_n) \preceq x \odot y$.

To prove $x \odot y \preceq s$ is a little harder. This inequality is, obviously, equivalent to the implication

$$x \in U \text{ and } y \in U \implies s \in U$$

holding for all ultrafilters U of $exp\, B$. By 3.3.2, each ultrafilter of $exp\, B$ is of the form $U(F)$ for some proper filter F of B. Decoding the definitions of $x, y, v(\ldots)$ and $U(F)$, we are left to prove that for all proper filters F of B

IF

(5) $\bigvee_{i=1}^{n}(e_i \wedge d_i) \in F$ and $\bigvee_{i=1}^{n}(c_i \wedge d_i) \in F$

AND

(6) for all i neither $-(e_i \wedge d_i) \in F$ nor $-(c_i \wedge d_i) \in F$

THEN

(7) $\bigvee_{i=1}^{n}(a_i \wedge c_i) \in F$ AND   (8) $-(a_i \wedge c_i) \notin F$ for all i.

To prove this implication, we invoke the taylormade formulas (1) and (2). From (5) and (1) we get

(9) $\bigvee_{i=1}^{n}(e_i \wedge c_i \wedge d_i) \in F$,

which together with (3) implies (7).

If (8) were false, then $-(a_j \wedge c_j) \leq -(e_j \wedge c_j) \in F$, for some j. Hence, by (9) and (2),

$$F \ni [\bigvee_{i=1}^{n}(e_i \wedge c_i \wedge d_i)] - (e_j \wedge c_j) \leq -(c_j \wedge d_j)$$

contradicting (6). Claim 4 is now proved and thereby the proposition. $\square$

For applications in section 5.4, we reformulate the above result.

Corollary 3.3.16 *If $\aleph_1$ is a caliber of A and $A \leq_{reg} B$, then $\pi(expA \,/\, expB) \leq \pi(A/B)$.*

This follows immediately from the proposition and the observation that all free powers $A^{(n)}$ have caliber $\aleph_1$ and, therefore, satisfy the ccc. We leave the easy verification as an exercise (use induction on n).

DIGRESSION

Here we sketch how exponentials can be introduced in the style in which we dealt with superextensions in the previous section. Let A be given. Denote by $E(A)$ the closed subspace of 2^A consisting of all $p : A \to 2$ such that

$$p(0) = 0, \quad p(1) = 1, \quad \text{and} \quad p(a \wedge b) = p(a) \wedge p(b).$$

Then put $exp A = Clop(E(A))$. The basic elements of $exp A$ are now defined by $v(a_1, \ldots, a_n) =$
$\{p \in E(A) : p(a_1 \vee \ldots \vee a_n) = 1 \text{ and } p((a_1 \vee \ldots \vee a_n) - a_i) = 0 \text{ for all } i\}$.
Lemmas 3.3.3 and 3.3.4 can be proved directly from this definition the latter with a counting argument again.

For a filter $F \subseteq A$ the mapping $p_F : A \to 2$ sending F to 1 and $A \setminus F$ to 0 is easily seen to belong to $E(A)$. Vice versa, each $p \in E(A)$ is of the form p_F for a suitable filter F. The correspondence between filters of A and ultrafilters of $exp A$ is therefore given by

$$U(F) = \{v \in Clop(E(A)) : p_F \in v\}.$$

On the base of this definition one can prove lemma 3.3.7. The rest of the above developpement then goes through as before.

3.4   Symmetric powers

The aim of this section is the construction of a relatively complete subalgebra of Fr ω_2 which is not projective. After exp Fr ω_2 this will be our second example of an rc-filtered, by 2.3.1, non-projective Boolean algebra. First we describe the example topologically. Let a topological space X and a natural number $n \geq 2$ be given. Define an equivalence relation $\sim$ on X^n by declaring two n-tuples equivalent iff one is a permutation of the other. The quotient space $X^n / \sim$ is called the n-th symmetric power of X and denoted by $SP^n(X)$. It is easy to check that the canonical projection $X^n \to SP^n(X)$ is an open mapping. In [53] Ščepin considers the Cantor cube 2^{ω_2} of weight $\aleph_2$ and proves that, for $n \geq 2$, all symmetric powers $SP^n(2^{\omega_2})$ are pairwise non-homeomorphic non-Dugundji spaces.[9]

It is part of this result that will be presented here. We confine ourselves to the case $n = 2$ and prove that the Boolean algebra dual to $SP^2(2^{\omega_2})$ is not projective.

As with exponentials we now give an algebraic description of the construction. Topology will play no explicit role in the rest of this section. Some familiarity with free products is required (cf. page 9).

Let a Boolean algebra A be given and consider the automorphism σ of $A \otimes A$ sending $a \otimes b$ to $b \otimes a$. An element of $A \otimes A$ will be called symmetric iff it is fixed by σ. By $SP^2(A)$ we denote the subalgebra of $A \otimes A$ consisting of all symmetric elements, symbolically

$(*)$ $SP^2(A) = \{h \in A \otimes A : \sigma(h) = h\}.$

To make SP^2 a functor we need to define it for homomorphisms $\varphi : A \to B$. Notice first that φ gives rise to a homomorphism $\bar{\varphi} : A \otimes A \longrightarrow B \otimes B$ determined by the formula $\bar{\varphi}(a_1 \otimes a_2) = \varphi(a_1) \otimes \varphi(a_2)$. An immediate verification shows that $\bar{\varphi} \circ \sigma_A = \sigma_B \circ \bar{\varphi}$ holds for all generators $a_1 \otimes a_2$ of $A \otimes A$. As there are homomorphisms on both sides, the same equation is true on all of $A \otimes A$, from which it follows that $\bar{\varphi}$ maps $SP^2(A)$ into $SP^2(B)$. So, we may define

$(**)$ $SP^2(\varphi) = \bar{\varphi} {\restriction} SP^2(A).$

We leave it to the reader to verify that conditions $(*)$ and $(**)$ define a functor to be called the *symmetric square*.

A final bit of notation is needed. For $a, b \in A$ we put $h(a, b) = (a \otimes b) \vee (b \otimes a)$. If there are several algebras to distinguish, we write h_A, h_B, etc. It is obvious that $h(a, b)$ belongs to $SP^2(A)$. The following formulas will be useful in the calculations below. Their straightforward verifications are left to the reader.

$(f1)$ $h(a, b) = h(b, a)$

$(f2)$ $h(a, b) = 0$ iff $a = 0$ or $b = 0$.

---

[9] In [53] the result is formulated for the functors exp_n but *proved* for SP^n. This was later noticed by Ščepin himself [54].

(f3) $h(a_1, b_1) \wedge h(a_2, b_2) = h(a_1 \wedge a_2, b_1 \wedge b_2) \vee h(a_1 \wedge b_2, a_2 \wedge b_1)$

(f4) $h(a_1 \vee a_2, b_1 \vee b_2) = h(a_1, b_1) \vee h(a_1, b_2) \vee h(a_2, b_1) \vee h(a_2, b_2)$

(f5) $a_1 \leq a_2$ and $b_1 \leq b_2$ imply $h(a_1, b_1) \leq h(a_2, b_2)$.

Here are some more

Observations:

(1) *Each element of $SP^2(A)$ can be written as a finite union of $h(a, b)$'s.*

Indeed, consider $c = \bigvee_{i=1}^n a_i \otimes b_i$. If c is symmetric, then

$$c = \sigma(c) = c \vee \sigma(c) = \bigvee_{i=1}^n a_i \otimes b_i \vee \bigvee_{i=1}^n b_i \otimes a_i = \bigvee_{i=1}^n h(a_i, b_i).$$

(2) $SP^2(A) \leq_{rc} A \otimes A$

Indeed, let $c \in A \otimes A$ be given. Then $c \leq c \vee \sigma(c) \in SP^2(A)$. If $c \leq d \in SP^2(A)$, then $\sigma(c) \leq \sigma(d) = d$, hence $c \vee \sigma(c) \leq d$, which shows that $c \vee \sigma(c)$ is minimal in $SP^2(A)$ above c.

(3) *If $\varphi : A \rightarrow B$ is a homomorphism, then*

$$SP^2(\varphi)(h_A(a, b)) = h_B(\varphi(a), \varphi(b)).$$

It follows that SP^2 preserves injectivity of homomorphisms. If $A \leq B$, then $SP^2(A)$ can be identified with the subalgebra of $SP^2(B)$ generated by all $h_B(a_1, a_2)$ with $a_1, a_2 \in A$.

This is obvious. From (1) and (3) it follows that

(4) SP^2 *is continuous and preserves cardinalities.*

As an immediate corollary to observation (2) and earlier theorems on free products, subalgebras and homomorphic images we get

Proposition 3.4.1 $SP^2(A)$ *is rc-filtered iff A is rc-filtered.*

Lemma 3.4.2 *Each ultrafilter of $SP^2(A)$ extends to at most two different ultrafilters of $A \otimes A$.*

Proof. Let $W \in Ult\, SP^2(A)$ be given and assume, by contradiction, that there were three different extensions to ultrafilters of $A \otimes A$. Then there exist elements $a_1 \otimes b_1, a_2 \otimes b_2$, and $a_3 \otimes b_3$ in $A \otimes A$ which are

(d) pairwise disjoint and

(e) such that $W \cup \{a_i \otimes b_i\}$ extends to an ultrafilter of $A \otimes A$ for $i = 1, 2, 3$.

From $a_i \otimes b_i \leq h(a_i, b_i) \in SP^2(A)$ and (e) we get $h(a_i, b_i) \in W$ for $i = 1, 2, 3$. But this is impossible, because, by $(f3)$,

$$h(a_1, b_1) \wedge h(a_2, b_2) \wedge h(a_3, b_3) =$$
$$h(a_1 \wedge a_2 \wedge a_3, b_1 \wedge b_2 \wedge b_3) \vee h(a_1 \wedge b_2 \wedge a_3, a_2 \wedge b_1 \wedge b_3)$$
$$\vee\, h(a_1 \wedge a_2 \wedge b_3, b_1 \wedge b_2 \wedge a_3) \vee h(a_1 \wedge b_2 \wedge b_3, a_2 \wedge b_1 \wedge a_3),$$

which is 0, by (d) and $(f2)$. $\Box$

Corollary 3.4.3 *If all ultrafilters of A have character κ, then all ultrafilters of $SP^2(A)$ also have character κ.*

Proof. Assume not. Then there is an ultrafilter W of $SP^2(A)$ which is generated by a subset W_0 of power $< \kappa$. The lemma yields an element $a \otimes b$ of $A \otimes A$ such that $W \cup \{a \otimes b\}$ generates an ultrafilter, U say, of $A \otimes A$. Clearly, $W_0 \cup \{a \otimes b\}$ also generates U, which contradicts the (folklore) fact that under the assumptions of the corollary all ultrafilters of $A \otimes A$ have character κ. $\Box$

The proof that $SP^2(\mathrm{Fr}\,\omega_2)$ is not projective splits into two parts. A 'skeleton chase' is used to produce some diagram and then a 'diagram argument' yields the desired contradiction. We hope to gain transparency if we separate the latter part beforehand and in abstract form. For a while we will be concerned with the following situation:

$A, B,$ and C *are atomless subalgebras of some Boolean algebra D and such that*

(g) $D = \langle A \cup B \cup C \rangle$ *and*

(i) $a \wedge b \wedge c \neq 0$ *for all non-zero $a \in A, b \in B,$ and $c \in C$.*

Let us introduce the abbreviations G for $SP^2(D)$ and E for the subalgebra of G generated by $SP^2(\langle A \cup B \rangle) \cup SP^2(\langle A \cup C \rangle)$.

What we need is the following

Lemma 3.4.4 *E is not relatively complete in G.*

Instead of proving this directly, we establisch the following two assertions.

(A) E is dense in G and (B) $E \neq G$.

From (A) and (B) the lemma follows easily (cf. the proof of 2.6.1).

Proof of (A). By observation (1) it is sufficient to find a non-zero $e \in E$ below any given non-zero $h(d_1, d_2) \in G$. From (g) we get that each $d \in D$ can be written in the form $\bigvee_{i=1}^{n} a_i \wedge b_i \wedge c_i$, with $a_i \in A, b_i \in B,$ and $c_i \in C$. Therefore, it will even suffice to find a non-zero element of E below any given

$$0 \neq h(a_1 \wedge b_1 \wedge c_1,\, a_2 \wedge b_2 \wedge c_2) \in G.$$

Using that A is atomless, we find non-zero $a' \leq a_1$ and $a'' \leq a_2$ such that $a' \wedge a'' = 0$. Then

$$0 \neq h(a' \wedge b_1 \wedge c_1,\, a'' \wedge b_2 \wedge c_2) \leq h(a_1 \wedge b_1 \wedge c_1,\, a_2 \wedge b_2 \wedge c_2),$$

by $(i), (f2)$, and $(f5)$. Moreover, by $(f3)$ and $(f2)$,

$$h(a' \wedge b_1 \wedge c_1, a'' \wedge b_2 \wedge c_2) = h(a' \wedge b_1, a'' \wedge b_2) \wedge h(a' \wedge c_1, a'' \wedge c_2)$$

belongs to E. □

Although probably more plausible than (A) assertion (B) takes more effort to prove. We shall use a general fact about Boolean algebras that we formulate beforehand as

Observation 3.4.5 *Let K and L be subsets of some Boolean algebra G. If they are downwards directed[10], then the set*

$$[K, L] = \{g \in G : \text{there are } k \in K, l \in l \text{ such that } k \vee l \leq g \text{ or } (k \vee l) \wedge g = 0\}$$

is a subalgebra of G.

Its straightforward verification is left to the reader. Still preparing the proof of (B) we notice that

Observation 3.4.6 *Every element of $SP^2(\langle A \cup B \rangle)$ can be written as a finite union of elements of the form $h(a_1 \wedge b_1, a_2 \wedge b_2)$ with $a_i \in A$ and $b_i \in B$.*

To check this, one needs observation (1) and formula $(f4)$.

Now we start the **proof** of (B) taking ultrafilters $U \in Ult\,A$, $V_1, V_2 \in Ult\,B$ and $W_1, W_2 \in Ult\,C$ such that $V_1 \neq V_2$ and $W_1 \neq W_2$. Then we put

$$K = \{h(a_1 \wedge b_1 \wedge c_1, a_2 \wedge b_2 \wedge c_2) : a_i \in U, \; b_i \in V_i \text{ and } c_i \in W_i\}$$
and
$$L = \{h(a_1 \wedge b_2 \wedge c_1, a_2 \wedge b_1 \wedge c_2) : a_i \in U, \; b_i \in V_i \text{ and } c_i \in W_i\}.$$

As ultrafilters are closed under meets, the formula

$$h(a'_1 \wedge a''_1 \wedge b'_1 \wedge b''_1 \wedge c'_1 \wedge c''_1, a'_2 \wedge a''_2 \wedge b'_2 \wedge b''_2 \wedge c'_2 \wedge c''_2) \leq$$
$$h(a'_1 \wedge b'_1 \wedge c'_1, a'_2 \wedge b'_2 \wedge c'_2) \wedge h(a''_1 \wedge b''_1 \wedge c''_1, a''_2 \wedge b''_2 \wedge c''_2)$$

shows that K is downwards directed. The same is true of L. Our strategy is to verify

$$(B1) \;\; E \subseteq [K, L] \quad \text{and} \quad (B2) \;\; [K, L] \neq G.$$

Because of the symmetry and the observations above, $(B1)$ reduces to showing that

$$h(a_1 \wedge b_1, a_2 \wedge b_2) \in [K, L] \quad \text{for all } a_1, a_2 \in A, \; b_1, b_2 \in B,$$

which is done by considering cases.
CASE 1. $a_i \notin U$, for $i = 1$ or 2.
Then $-a_i \in U$, hence $h(-a_i, -a_i) \in K \cap L$. Moreover, by $(f2)$ and $(f3)$,

$$h(-a_i, -a_i) \wedge h(a_1 \wedge b_1, a_2 \wedge b_2) = 0.$$

[10]i.e. if for all $k_1, k_2 \in K$ there is some $k \in K$ such that $k \leq k_1 \wedge k_2$ and the same with L.

CASE 2. $a_1, a_2 \in U$, $b_1, b_2 \notin V_i$, for $i = 1$ or 2.
Taking $b \in V_i$ such that $b \wedge (b_1 \vee b_2) = 0$, we get $h(b, 1) \in K \cap L$ and

$$h(b, 1) \wedge h(a_1 \wedge b_1, a_2 \wedge b_2) = 0.$$

CASE 3. $a_1, a_2 \in U$, $b_1 \in V_1$, $b_2 \in V_2$.
Putting $a = a_1 \wedge a_2$ we have $h(a \wedge b_1, a \wedge b_2) = h(a \wedge b_2, a \wedge b_1) \in K \cap L$ and

$$h(a \wedge b_1, a \wedge b_2) \leq h(a_1 \wedge b_1, a_2 \wedge b_2).$$

CASE 4. $a_1, a_2 \in U$, $b_2 \in V_1$, $b_1 \in V_2$ is actually the same as case 3. So $(B1)$ is proved.

We prove $(B2)$ by exhibiting an element of G which is not in $[K, L]$. Let us take elements $b' \in V_1$, $b'' \in V_2$, $c' \in W_1$, $c'' \in W_2$ such that $b' \wedge b'' = c' \wedge c'' = 0$.
    Then $h(b' \wedge c', b'' \wedge c'')$ intersects every element $h(a_1 \wedge b_1 \wedge c_1, a_2 \wedge b_2 \wedge c_2) \in K$. Indeed, from

$$b' \wedge b_1 \neq 0, \ b'' \wedge b_2 \neq 0, \ c' \wedge c_1 \neq 0, \ \text{and} \ c'' \wedge c_2 \neq 0$$

(because the corresponding elements belong to the same ultrafilter) we get, by (i) and $(f2)$,

$$\begin{aligned} 0 \neq \ & h(a_1 \wedge b_1 \wedge b' \wedge c_1 \wedge c', a_2 \wedge b_2 \wedge b''_j \wedge c_2 \wedge c'') \\ \leq \ & h(a_1 \wedge b_1 \wedge c_1, a_2 \wedge b_2 \wedge c_2) \wedge h(b' \wedge c', b'' \wedge c''). \end{aligned}$$

On the other hand $h(b' \wedge c', b'' \wedge c'')$ contains no element of L. Indeed, assume

$$h(a_1 \wedge b_2 \wedge c_1, a_2 \wedge b_1 \wedge c_2) \leq h(b' \wedge c', b'' \wedge c'').$$

Replacing b_1 by $b_1 \wedge b'$ and b_2 by $b_2 \wedge b''$ if necessary, we can further assume $b_1 \leq b'$ and $b_2 \leq b''$. Hence, by $(f2)$ and $(f3)$,

$$h(a_1 \wedge b_2 \wedge c_1, a_2 \wedge b_1 \wedge c_2) = h(a_1 \wedge b_2 \wedge c_1, a_2 \wedge b_1 \wedge c_2) \wedge h(b' \wedge c', b'' \wedge c'') = 0,$$

contradicting (i).
    It follows that $h(b' \wedge c', b'' \wedge c'') \notin [K, L]$ and assertion (B) is finally proved. Thereby, also lemma 3.4.4. $\square$

Now we are ready to prove the main result of this section.

Theorem 3.4.7 (Ščepin [53]) $SP^2(\text{Fr } \omega_2)$ *is not projective.*

Proof. For a subset X of ω_2 we write $S(X)$ to denote the subalgebra $SP^2(\text{Fr } X)$ of $SP^2(\text{Fr } \omega_2)$. Assume, by contradiction, that $S(\omega_2)$ were projective. By lemma 3.4.3 all ultrafilters of $S(\omega_2)$ have character $\aleph_2$. So, Ščepin's theorem 1.4.11 yields an isomorphism $\varphi : S(\omega_2) \longrightarrow \text{Fr } \omega_2$. A standard argument shows that the set of all $X \subseteq \omega_2$ such that $|X| = \aleph_1$ and $\varphi(S(X)) = \text{Fr } X$ is unbounded in the partially ordered (under $\subseteq$) set $[\omega_2]^{\aleph_1}$. This fact allows us to choose disjoint subsets X, Y of ω_2 both of power $\aleph_1$ such that

$$(*)\qquad \varphi(S(X)) = \mathrm{Fr}\, X \quad \text{and} \quad \varphi(S(X \cup Y)) = \mathrm{Fr}\,(X \cup Y).$$

Now the same kind of argument is repeated in a slightly refined way. Let us consider the partial order $\mathcal{P}$ consisting of all pairs (X', Y'), where X' and Y' are countable subsets of X and Y, respectively, ordered by inclusion, i.e.

$$(X', Y') \leq (X'', Y'') \Longleftrightarrow X' \subseteq X'' \quad \text{and} \quad Y' \subseteq Y''.$$

CLAIM   *The set $\mathcal{Q}$ of all $(X', Y') \in \mathcal{P}$ such that* $\varphi(S(X')) = \mathrm{Fr}\, X'$ *and* $\varphi(S(X' \cup Y')) = \mathrm{Fr}\,(X' \cup Y')$ *is unbounded in $\mathcal{P}$.*

Indeed, consider an arbitrary $(X_0, Y_0) \in \mathcal{P}$. Use $(*)$ and the continuity of S and Fr to find $(X_0, Y_0) \leq (X_1, Y_1) \in \mathcal{P}$ such that

$$\varphi(S(X_0)) \subseteq \mathrm{Fr}\, X_1 \quad \text{and} \quad \varphi(S(X_0 \cup Y_0)) \subseteq \mathrm{Fr}\,(X_1 \cup Y_1).$$

Then use $(*)$ again to find $(X_1, Y_1) \leq (X_2, Y_2) \in \mathcal{P}$ such that

$$\mathrm{Fr}\, X_1 \subseteq \varphi(S(X_2)) \quad \text{and} \quad \mathrm{Fr}\,(X_1 \cup Y_1) \subseteq \varphi(S(X_2 \cup Y_2)).$$

Proceeding that way one gets a sequence $(X_0, Y_0) \leq (X_1, Y_1) \leq (X_2, Y_2) \leq \cdots$ such that $(\bigcup_{n < \omega} X_n, \bigcup_{n < \omega} Y_n)$ belongs to $\mathcal{Q}$. This proves the claim.

It follows from the claim that there are pairwise disjoint countably infinite sets $X_0, X_1 \subseteq X$ and $Y_0, Y_1 \subseteq Y$ such that the diagram

$$
\begin{array}{ccc}
S(X_0 \cup X_1) & \leq & S(X_0 \cup X_1 \cup Y_0 \cup Y_1) \\
\mathsf{VI} & & \mathsf{VI} \\
S(X_0) & \leq & S(X_0 \cup Y_0)
\end{array}
$$

is isomorphic under φ to the diagram

$$
\begin{array}{ccc}
\mathrm{Fr}\,(X_0 \cup X_1) & \leq & \mathrm{Fr}\,(X_0 \cup X_1 \cup Y_0 \cup Y_1) \\
\mathsf{VI} & & \mathsf{VI} \\
\mathrm{Fr}\,(X_0) & \leq & \mathrm{Fr}\,(X_0 \cup Y_0).
\end{array}
$$

Clearly, $\mathrm{Fr}\,(X_0 \cup X_1) \cup \mathrm{Fr}\,(X_0 \cup Y_0)$ generates $\mathrm{Fr}\,(X_0 \cup X_1 \cup Y_0)$, which is a relatively complete subalgebra of $\mathrm{Fr}\,(X_0 \cup X_1 \cup Y_0 \cup Y_1)$. It follows that $S(X_0 \cup X_1) \cup S(X_0 \cup Y_0)$ generates a relatively complete subalgebra of $S(X_0 \cup X_1 \cup Y_0 \cup Y_1)$. As $S(X_0 \cup X_1)$ and $S(X_0 \cup Y_0)$ are both contained in $S(X_0 \cup X_1 \cup Y_0)$ we must have

$$\langle S(X_0 \cup X_1) \cup S(X_0 \cup Y_0) \rangle \leq_{rc} S(X_0 \cup X_1 \cup Y_0),$$

by transitivity of relative completeness. But this contradicts lemma 3.4.4 (with $A = \mathrm{Fr}\, X_0, B = \mathrm{Fr}\, X_1$ and $C = \mathrm{Fr}\, Y_0$). $\square$

From the above theorem we get.

Corollary 3.4.8 $SP^2(A)$ *is projective iff A is projective and $|A| \leq \aleph_1$.*

Proof.   The homomorphism $(id \otimes id) \restriction SP^2(A) : SP^2(A) \longrightarrow A$ sends $h(a, b)$ to $a \wedge b$. Indeed, as $id \otimes id$ sends $a \otimes b$ to $a \wedge b$, we get

$$(id \otimes id)(h(a, b)) = (a \wedge b) \vee (b \wedge a) = a \wedge b.$$

The mapping $a \mapsto h(a, a) = a \otimes a$ is, obviously, a right inverse of $id \otimes id$. As it preserves 0 and $\wedge$, the projectivity of $SP^2(A)$ implies that of A, by 2.4.3.

The rest is proved follows exactly as for exp (in 3.3.10). $\square$

Chapter 4

σ-filtered Boolean algebras

In this chapter we take a short look at the class of Boolean algebras that is adequate to the class of σ-embeddings. We start with a concept that will seem unrelated at first sight.

4.1 Bicommutative diagrams and commuting subalgebras

In the topological theory of open generated spaces so-called bicommutative diagrams play an important role. A commutative quadratic diagram

$$\begin{array}{ccc} X & \xrightarrow{f} & Y \\ h \downarrow & & \downarrow g \\ T & \xrightarrow{k} & Z \end{array}$$

is called *bicommutative* if one of the following equivalent conditions holds

(1) $f(h^{-1}(t)) = g^{-1}(k(t))$, for all $t \in T$.

(2) $h(f^{-1}(y)) = k^{-1}(g(y))$, for all $y \in Y$.

(3) If $g(y) = k(t)$, then $y = f(x)$ and $t = h(x)$, for some $x \in X$.

To be definite, we take clause (3) as definition and leave the equivalence proofs to the interested reader (cf. [40], section 3).

In this section we explain how this concept translates into Boolean algebra and why we did not explicitly need it in our developement. If X, Y, Z, T are Boolean spaces, the above diagram gives rise to a diagram of embeddings of Boolean algebras which we call (∗) and fix for a while.

$$
\begin{array}{ccc}
A & \leq & D \\
& \text{\Large V/} & \text{\Large V/} \\
C & \leq & B
\end{array}
$$

$(*)$

Definition 4.1.1 We say that A and B *commute over* C and write $A \downarrow_C B$ if one of the following conditions is satisfied for all $a \in A, b \in B$.

($\downarrow$ 1) If $a \wedge b = 0$, then $a \leq c$ and $b \leq d$, for some disjoint $c, d \in C$.

($\downarrow$ 2) If $a \leq b$, then $a \leq c \leq b$, for some $c \in C$.

If $A \leq D \geq B$, we say that A and B commute $(A \downarrow B)$ if they commute over $A \cap B$. □

This definition is a generalization of the usual notion of independence of sub-algebras (mentioned in section 1.4, see also [34], 11.3) which corresponds to commutativity over $\{0, 1\}$. So it is natural to characterize commutativity as a freedom-like property. In the situation $(*)$ we introduce two more conditions.

($\downarrow$ 3) For all Boolean algebras E and all pairs of homomorphisms $\varphi : A \to E$ and $\psi : B \to E$ such that $\varphi \restriction C = \psi \restriction C$ there exists a homomorphism $\theta : \langle A \cup B \rangle \to E$ such that $\varphi \cup \psi \subseteq \theta$.

($\downarrow$ 4) For all $U \in Ult A$ and $V \in Ult B$ such that $U \cap C = V \cap C$ there is some $W \in Ult D$ such that $U \cup V \subseteq W$.

It is an easy exercise in Stone duality to verify that condition ($\downarrow$4) is equivalent to the bicommutativity of the diagram

$$
\begin{array}{ccc}
Ult D & \longrightarrow & Ult A \\
\downarrow & & \downarrow \\
Ult B & \longrightarrow & Ult C
\end{array}
$$

dual to $(*)$, where all arrows are the mappings sending ultrafilters of the big algebra to their intersection with the smaller one.

Proposition 4.1.2 *In the situation $(*)$ all conditions ($\downarrow$1) – ($\downarrow$4) are equivalent.*

Proof. ($\downarrow$ 1) $\Longleftrightarrow$ ($\downarrow$ 2) is easy. After the identification of ultrafilters with homomorphisms into the two-element Boolean algebra, the implication ($\downarrow$3) $\Longrightarrow$ ($\downarrow$4) should also be clear: ($\downarrow$3) yields an extension of $U \cup V$ to an ultrafilter of $\langle A \cup B \rangle$, which is then easily extended to the desired W.

In order to prove ($\downarrow$4) $\Longrightarrow$ ($\downarrow$1), assume $a \wedge b = 0$. Consider $F = \{c \in C : c \geq a\}$ and $G = \{c \in C : c \geq b\}$. Both sets are filters of C. It will be sufficient to show that $F \cup G$ does not have the finite intersection property. To get a contradiction, we assume it does and take some $R \in Ult C$ extending $F \cup G$. Then $R \cup \{a\}$ also has the finite intersection property. For, otherwise, there were some $e \in R$ such that $e \wedge a = 0$, i.e. $a \leq -e$, from which $-e \in F \subseteq R$ contradicting $e \in R$.

Let us take $U \in UltA$ containing R and a. By symmetry, there is some $V \in UltB$ containing R and b. As $U \cap C = V \cap C = R$, condition ($\downarrow$4) yields some $W \in UltD$ containing $U \cup V$. But then $a, b \in W$, which is impossible.

To prove ($\downarrow$1) $\Longrightarrow$ ($\downarrow$3) we apply Sikorski's Extension Criterion (cf. 0.0.1) to the mapping $\varphi \cup \psi : A \cup B \longrightarrow E$. It boils down to the condition

$$\forall a \in A \, \forall b \in B \quad a \wedge b = 0 \implies \varphi(a) \wedge \psi(b) = 0.$$

But if $a \wedge b = 0$, then, by ($\downarrow$1), $a \leq c, b \leq d$, for some disjoint $c, d \in C$. Hence

$$\varphi(a) \wedge \psi(b) \leq \varphi(c) \wedge \psi(d) = \varphi(c) \wedge \varphi(d) = 0.$$

So, Sikorski's criterion applies and the proposition is proved. □

The following observations give examples of what the notion of commutativity can be useful for. The first one refers to the situation $(*)$, i.e.

$$
\begin{array}{ccc}
A & \leq & D \\
\vee/ & & \vee/ \\
C & \leq & B
\end{array}
$$

Observation 4.1.3 *If $A \downarrow_C B$ and $A \leq_{rc} D$, then $C \leq_{rc} B$ and $p_C^B = p_A^D \restriction B$.*

Recall that p_C^B denotes the lower projection. The same assertion is true for the upper projection q_C^B, of course.

Proof. Take $b \in B$ and put $a = p_A^D(b)$. We show that $a \in C$. From $a \leq b$ and $A \downarrow_C B$ we get some $c \in C$ such that $a \leq c \leq b$. But $C \subseteq A$ and a was maximal in A below b. So $c \leq a$. □

Observation 4.1.4 *Assume $A \leq B$ and let $(B_i)_{i \in I}$ be a filtration of B. If $A \cap B_i \leq_{rc} B_i$ for all $i \in I$ and $A \cap B_j \downarrow B_i$ for all $i \leq j$, then $A \leq_{rc} B$.*

Proof. Let $b \in B$ be given. Then $b \in B_i$ for some i. Take $a \in A \cap B_i \leq_{rc} B_i$ maximal below b. To show that a is maximal in A below b, we consider any $a' \in A$ such that $a' \leq b$. As I is directed, $a' \in A \cap B_j$ for some $j \geq i$. As $A \cap B_j$ and B_i commute, we find $a'' \in A \cap B_j \cap B_i = A \cap B_i$ such that $a' \leq a'' \leq b$. But then $a'' \leq a$, because a was maximal in $A \cap B_i$. So $a' \leq a$ and we are done. □

Now we show what we replaced the explicit use of commutativity with.

Proposition 4.1.5 *Assume $A \leq_{rc} D \geq B$. Then $A \downarrow B$ iff B is closed under the projection p_A^D.*

Proof. Denote p_A^D by p. Assume first $A \downarrow B$ and consider $b \in B$. Then $A \ni p(b) \leq b$, hence $p(b) \leq c \leq b$ for some $c \in A \cap B$. But $p(b)$ is maximal in A below b, hence $c \leq p(b)$. So, $p(b) = c \in B$, as desired.

Assume now that $p(B) \subseteq B$ and consider $a \leq b$. Then $a \leq p(b) \leq b$, and $p(b) \in A \cap B$. □

From the last proposition it follows that for each $B \leq_{rc} A$ there is a skeleton of subalgebras of A all members of which commute with B – just take the fixed-point skeleton of the projection. Next we prove a generalization of this fact. Recall that we defined $A \leq B$ to be a σ-embedding (in symbols $A \leq_\sigma B$) if $A \cap I$ is countably generated whenever I is a countably generated ideal of B. For that to hold it is sufficient, of course, that $A \restriction b$ is countably generated for all $b \in B$.

Proposition 4.1.6 ([55] Theorem 2.2)
$A \leq_\sigma B$ iff $\{C \leq B : A \downarrow C\}$ *is a skeleton of* B.

Proof. Assume $A \leq_\sigma B$. Let $f : B \longrightarrow [A]^{\leq \aleph_0}$ be such that $f(b)$ is closed under finite joins and generates $A \restriction b$. Let $S \leq B$ be closed under f. We show that $A \downarrow S$. Suppose $a \leq s$ for $a \in A$ and $s \in S$. Then $a \in A \restriction s$ and there is some $b \in f(s) \subseteq A \cap S$ such that $a \leq b \leq s$. This b witnesses ($\downarrow 2$).

It follows that $\{C \leq B : A \downarrow C\}$ contains the fixed-point skeleton of f and is, therefore, absorbing. A moments reflection shows that it is also closed.

Assume now that $A \leq B$ commutes with all algebras of some skeleton $\mathcal{S}$. Let b in B be given. Find a countable $S \in \mathcal{S}$ such that $b \in S$. Then $(A \cap S) \restriction b$ generates $A \restriction b$. Indeed, as $A \downarrow S$, we find for all $a \leq b$, some $s \in S \cap A$ such that $a \leq s \leq b$. As $|A \cap S| \leq |S| = \aleph_0$, we are done. □

We call a family consisting of pairwise commuting subalgebras a *commutative family*. Then the above result immediately yields

Corollary 4.1.7 *Any commutative skeleton consists of σ-embedded subalgebras.*

4.2 The weak Freese/Nation property

In this section we prove a converse to the last corollary, namely *any skeleton consisting of σ-embedded subalgebras contains a commutative subskeleton.* To give a foretaste, let us prove first that *rc-filtered Boolean algebras have commutative skeletons.* We use that rc-filtered algebras have the Freese/Nation property. Let $I : A \rightarrow [A]^{< \aleph_0}$ witness (FNI) for A. Then the fixed-point skeleton of I is as desired. Indeed, if $S, T \leq A$ are closed under I, then $S \downarrow T$. For, $s \leq t$ implies $s \leq x \leq t$ for some $x \in I(s) \cap I(t) \subseteq S \cap T$.

In this argument the finiteness of the $I(a)$ is not essential. What we need is a fixed-point skeleton and that exists also if all $I(a)$ are only countable. This observation motivates the following

Definition 4.2.1 We say that a Boolean algebra A has the *weak Freese/Nation property*, abbreviated as (WFN), if there is a mapping $I : A \rightarrow [A]^{\leq \aleph_0}$ such that $a \leq b$ implies $a \leq x \leq b$ for some $x \in I(a) \cap I(b)$. □

As in the finite case, there is a separation version.

Observation 4.2.2 *The Boolean algebra A has the property (WFN) iff there is a mapping $S : A \to [A]^{\leq \aleph_0}$ such that $a \wedge b = 0$ implies $a \leq c$ and $b \leq d$ for some disjoint $c, d \in S(a) \cap S(b)$.*

Of either I or S we shall say that they witness (WFN) for A.

Now we have everything together to formulate the main result of this section.

Theorem 4.2.3 *The following assertions are equivalent for an arbitrary Boolean algebra A.*

  (1) *A has the property (WFN).*

  (2) *A has a commutative skeleton.*

  (3) *A has a skeleton consisting of σ-embedded subalgebras.*

  (4) *A has a commutative club.*

  (5) *There is a commutative family $\mathcal{C}$ of countable subalgebras of A such that $A = \bigcup \mathcal{C}$.*

Proof. $(1) \Rightarrow (2)$ was sketched before the definition of (WFN).

$(2) \Rightarrow (4)$. The club consists of the countable members of the skeleton.

$(4) \Rightarrow (5)$. The club is such a family.

$(5) \Rightarrow (1)$. For each $a \in A$ let $I(a)$ be a member of $\mathcal{C}$ such that $a \in I(a)$. Then I witnesses (WFN).

It remains to incorporate (3). One direction is done already: $(2) \Rightarrow (3)$ follows from 4.1.7. The proof of $(3) \Rightarrow (1)$ will be based on the following result that is interesting in its own right.

Proposition 4.2.4 (Theorem 2.4 in [55]) *If A has a well-ordered filtration $(A_\alpha)_{\alpha < \lambda}$ such that all A_α are σ-embedded in A and have the property (WFN), then A also has this property.*

Let us assume for a moment that we have proved the proposition and derive the implication $(3) \Rightarrow (1)$ from it.

Let S be a skeleton of A such that $S \leq_\sigma A$ for all $S \in \mathcal{S}$. Let $\mathcal{C}$ denote the set of all countable members of $\mathcal{S}$. Clearly, $A = \bigcup \mathcal{C}$ and $\mathcal{C}$ is directed under $\subseteq$. So we will be done if we prove the following

CLAIM *If $\mathcal{D} \subseteq \mathcal{C}$ is directed, then $\bigcup \mathcal{D}$ has the property (WFN).*

Notice first that directedness implies that $\bigcup \mathcal{D}$ is a subalgebra of A. We prove the claim by induction on the cardinality of $\mathcal{D}$. If it is finite or countable, then $\bigcup \mathcal{D}$ is also countable and (WFN) is trivially true. Otherwise, we use Iwamura's Lemma 2.1.6 to decompose $\mathcal{D} = \bigcup_{\alpha < \lambda} \mathcal{D}_\alpha$ into a well-ordered continuous chain of directed subsets of smaller cardinality. Clearly, $(\bigcup \mathcal{D}_\alpha)_{\alpha < \lambda}$ is a well-ordered filtration of $\bigcup \mathcal{D}$. By 2.1.8, each $\bigcup \mathcal{D}_\alpha$ belongs to $\mathcal{S}$ and is, therefore, σ-embedded in A, hence in $\bigcup \mathcal{D}$. By induction hypothesis, each $\bigcup \mathcal{D}_\alpha$ has (WFN). So the proposition allows us to conclude that $\bigcup \mathcal{D}$ also has (WFN), as desired. $\square$

Now we **prove** the proposition. Its statement is analogous to that of 2.2.4 and so will be its proof.

Let $a \in A$ and $\alpha < \lambda$ be given. Denote the filter $\{b \in A_\alpha : a \leq b\}$ by $F_\alpha(a)$. Notice that, for every fixed a, the sets $F_\alpha(a)$ form a countinuously increasing chain of subsets of A, which eventually (as soon as $a \in A_\alpha$) contain a.

From $A_\alpha \leq_\sigma A$ it follows that we can take a countable set $Q_\alpha(a)$ which is closed under intersections and generates the filter $F_\alpha(a)$. In other words, $Q_\alpha(a)$ is a countable dense subset of $F_\alpha(a)$.

Call α a *jump point* of a if $F_\alpha(a)$ is not dense in $F_{\alpha+1}(a)$. Let $D(a)$ denote the set of all jump points of a.

CLAIM 1. *$D(a)$ is countable for each $a \in A$.*

Assume not. Then there is a subset, W say, of $D(a) \subseteq \lambda$ which has order type ω_1. Put $\gamma = \sup W$, the supremum being taken in $\lambda + 1$. (To be on the safe side we put $A_\lambda = A$ and $Q_\lambda(a) = \{a\}$.) As $Q_\gamma(a)$ is a countable subset of $F_\gamma(a) = \bigcup_{\alpha < \gamma} F_\alpha(a) = \bigcup_{\delta \in W} F_\delta(a)$ and $cf(W) = \omega_1$, there must be some $\delta \in W$ such that $Q_\gamma(a) \subseteq F_\delta(a)$. From

$$Q_\gamma(a) \subseteq F_\delta(a) \subseteq F_{\delta+1}(a) \subseteq F_\gamma(a)$$

and the density of Q_γ in F_γ it follows that $F_\delta(a)$ must be dense in $F_{\delta+1}(a)$, which contradicts $\delta \in W \subseteq D(a)$ and proves claim 1.

Let us write $U \wedge V$ for $\{u \wedge v : u \in U, v \in V\}$.

CLAIM 2. *If $a_1 \wedge a_2 = 0$, then either $0 \in F_0(a_1) \wedge F_0(a_2)$ or there is some $\alpha \in D(a_1) \cap D(a_2)$ such that $0 \in F_{\alpha+1}(a_1) \wedge F_{\alpha+1}(a_2)$.*

If β is big enough, then $0 = a_1 \wedge a_2 \in F_\beta(a_1) \wedge F_\beta(a_2)$. Consider the least β with this property. If $\beta = 0$, then we are done. Next we show that β cannot be a limit ordinal. Suppose it is. Then, as the sequences F_α increase continuously,

$$0 \in F_\beta(a_1) \wedge F_\beta(a_2) = \bigcup_{\alpha < \beta} [F_\alpha(a_1) \wedge F_\alpha(a_2)],$$

hence $0 \in F_\alpha(a_1) \wedge F_\alpha(a_2)$ for some $\alpha < \beta$, contradicting the minimality of β.

It remains to show that $\beta = \alpha + 1$ implies $\alpha \in D(a_1) \cap D(a_2)$. Let $b_1 \in F_{\alpha+1}(a_1)$ and $b_2 \in F_{\alpha+1}(a_2)$ be such that $b_1 \wedge b_2 = 0$ and, to get a contradiction, assume $\alpha \notin D(a_1)$. Then $F_\alpha(a_1)$ is dense in $F_{\alpha+1}(a_1)$, so we can find $c_1 \in F_\alpha(a_1)$ such that $c_1 \leq b_1$. Then $c_1 \wedge a_2 \leq b_1 \wedge b_2 = 0$ implies $a_2 \leq -c_1 \in A_\alpha$, i.e. $-c_1 \in F_\alpha(a_2)$. So $0 = c_1 \wedge -c_1 \in F_\alpha(a_1) \wedge F_\alpha(a_2)$, again contradicting the minimality of $\beta = \alpha + 1$.

Claim 2 is also proved. Together with the fact that each Q is dense in the corresponding F it yields

CLAIM 3. Claim 2 *remains true if F is replaced by Q*.

Now it is easy to finish the proof of the proposition. Let S_α witness (the separation version of) (WFN) for A_α and put

$$S(a) = \bigcup \{S_\beta(b) : b \in Q_\beta(a), \; \beta = 0 \text{ or } \beta - 1 \in D(a)\}.$$

By claim 1, each $S(a)$ is countable. If $a_1 \wedge a_2 = 0$, then, by claim 3, $b_1 \wedge b_2 = 0$ for some $b_1 \in Q_\beta(a_1)$, $b_2 \in Q_\beta(a_2)$ with $\beta = 0$ or $\beta - 1 \in D(a_1) \cap D(a_2)$. Then there are some disjoint $c_1, c_2 \in S_\beta(b_1) \cap S_\beta(b_2) \subseteq S(a_1) \cap S(a_2)$ such that $b_1 \le c_1$ and $b_2 \le c_2$. From $b_1 \in F_\beta(a_1)$ we get $a_1 \le b_1$, hence $a_1 \le c_1$. Analogously, $a_2 \le c_2$, as desired. $\square$

Theorem 4.2.3 is now completely proved. As in earlier chapters we take the condition on skeletons as basic for the following

Definition 4.2.5 A Boolean algebra will be called σ-*filtered* if it has a σ-skeleton, i.e. one that consists of σ-embedded subalgebras.

From the above theorem we immediately get

Corollary 4.2.6 *All Boolean algebras of cardinality at most $\aleph_1$ are σ-filtered.*

Indeed, just decompose the given algebra into a *chain* of countable subalgebras and use 4.2.3(5). $\square$

For example, under (CH), the power-set algebra on ω is σ-filtered. We don't know[1] whether one can prove this in ZFC.

The corollary shows that the (BSP) need not be true of σ-filtered algebras and also that there can be uncountable chains and uncountable sets of pairwise disjoint elements. But their size is limited, for

Proposition 4.2.7 *If A is σ-filtered, then $c(A) \le 2^{\aleph_0}$ and all subchains of A have cardinality at most $2^{\aleph_0}$, i.e. Length$(A) \le 2^{\aleph_0}$.*

Proof. Let $S : A \to [A]^{\le \aleph_0}$ witness the separation version of (WFN) and assume that $D \subseteq A$ consist of pairwise disjoint elements where $|D| > 2^{\aleph_0}$. Then $\{S(d) : d \in D\}$ is a collection of more than $2^{\aleph_0}$ countable sets, to which the Δ-Lemma (cf. 0.0.2) applies: there is a subset D' still of power $> 2^{\aleph_0}$ such that $S(d') \cap S(d'') = M$ for all distinct $d', d'' \in D'$. Using that S witnesses the separation version of (WFN) it is easy to verify that $d' \mapsto \{x \in M : d' \le x\}$ is an injective mapping $D' \to P(M)$. This implies $2^{\aleph_0} < |D'| \le 2^{|M|} \le 2^{\aleph_0}$, which is a contradiction proving $c(A) \le 2^{\aleph_0}$.

The assertion concerning chains is proved in exactly the same way, using I instead of S. $\square$

To see that the above results are best possible we describe an

Example. Let R be the set of real numbers with their usual order. For a subset

---

[1] See note added in proof at the end of this section.

L of R we let $B(L)$ denote the subalgebra of $Pow(R)$ generated by all open intervals (a, b) with endpoints $a, b \in L \cup \{+\infty, -\infty\}$.

As all singletons $\{x\} = R \setminus [(-\infty, x) \cup (x, \infty)]$ belong to $B(R)$, we get $c(B(R)) = 2^{\aleph_0}$. The chain $\{(-\infty, x) : x \in R\}$ shows $Length(B(R)) = 2^{\aleph_0}$.

It remains to see that $B(R)$ is σ-filtered. Leaving most of the verifications to the reader, we outline how to proceed. To make the following calculations easier one first establishes

CLAIM 1. *The elements of $B(L)$ are finite unions of pairwise disjoint sets that are either open intervals with endpoints in $L \cup \{+\infty, -\infty\}$ or singletons contained in L.*

Then one proves the decisive

CLAIM 2. *If L is dense in R, then $B(L) \leq_\sigma B(R)$.*

We just consider one case, which brings out the idea. Let a be in L and $b \in (R \setminus L) \cup \{+\infty\}$. We show that $B(L){\restriction}(a, b)$ is countably generated. Indeed, as L is dense in R, there is an increasing sequence $(l_n)_{n<\omega}$ of elements of L converging to b. Then, using claim 1, the countable set $\{(a, l_n) : n < \omega\} \subseteq B(L)$ is easily checked to generate $B(L){\restriction}(a, b)$.

From claim 2 it finally follows that

$$\{B(L) : L \text{ a dense subset of } R\}$$

is a skeleton of $B(R)$ consisting of σ-embedded subalgebras. □

At the moment we cannot prove the following

Conjecture [2] *If A is σ-filtered, then $Depth(A) \leq \aleph_1$ and $|A| \leq 2^{ind(A)}$.*

We end this section by discussing some more (because of 4.2.4) closedness properties. Concerning subalgebras we have

Proposition 4.2.8 *If $A \leq_\sigma B$ and B is σ-filtered, then A is also σ-filtered.*

The **proof** is a modification of that of 2.3.1. Instead of the projection we now use a mapping $P : B \to [A]^{\leq \aleph_0}$ such that $P(b)$ is closed under finite joins and generates $A{\restriction}b$. Then one puts $S_A(a) = \bigcup\{P(b) : b \in S_B(a)\}$ and verifies that this works. □

To characterize the adequate homomorphisms we introduce the following mode of speech.

Definition 4.2.9 A homomorphism $\varphi : A \to B$ will be called *approximately ⊥-invertible* if there is a sequence $\varepsilon = (\varepsilon_n)_{n<\omega}$ of mappings $\varepsilon_n : B \to A$ such that

---

[2] See note added in proof at the end of this section.

(i) $\varphi \circ \varepsilon_n = id_B$, for all n,

(ii) $\varepsilon_n(b) \geq \varepsilon_{n+1}(b)$, for all b and n, and

(iii) if $b_1 \wedge b_2 = 0$, then $\varepsilon_n(b_1) \wedge \varepsilon_n(b_2) = 0$, for almost all n. $\square$

Intuitively, $\varepsilon_\infty = \lim_{n \to \infty} \varepsilon_n$ (which usually does not exist) is a $\perp$-inverse of φ.

Proposition 4.2.10 *Let* $\varphi : A \to B$ *be a surjective homomorphism.*

(1) *If* A *is* σ-*filtered and* φ *is approximately* $\perp$-*invertible, then* B *is also* σ-*filtered.*

(2) *If* B *is* σ-*filtered, then* φ *is approximately* $\perp$-*invertible.*

The **proof** is a modification of the proof of 2.4.5. We confine ourselves to assertion (2). Let S witness (WFN) for B. For each $b \in B$ we decompose $S(b) = \bigcup_{n < \omega} S_n(b)$ into an increasing chain of finite subsets. As in the earlier proof we choose a mapping $\delta : B \to A$ such that $\varphi \circ \delta = id_B$ and $\delta(-b) = -\delta(b)$, for all b. Then we put

$$\varepsilon_n(b) = \delta(b) \wedge \bigwedge\{\delta(c) : b \leq c \text{ and } c \in S_n(b) \text{ or } -c \in S_n(b)\}.$$

Conditions (i) and (ii) are more or less obvious and (iii) is checked as in the old proof. $\square$

There should be an analogous proposition for approximatly $\leq$-invertible homomorphisms. Probably one has to give up condition (ii) for that purpose. Unfortunately we were only able to prove the easy direction of the corresponding proposition. As an application of 4.2.10 we draw the following

Corollary 4.2.11 *If* I *is a countably generated ideal of the* σ-*filtered Boolean algebra* A, *then* A/I *is also* σ-*filtered.*

Proof. We check that the canonical projection $\pi : A \to A/I$ is approximately $\perp$-invertible. Let $\{i_n : n < \omega\}$ be a set of generators of I and put $a_n = \bigvee_{m=0}^n i_m$.

By surjectivity of the projection, there exists a right inverse $\delta : A/I \to A$ of π. We define $\varepsilon_n : A/I \to A$ by $\varepsilon_n(b) = \delta(b) - a_n$. A straightforward verification then shows that $\varepsilon = (\varepsilon_n)_{n < \omega}$ is as desired. $\square$

Note added in proof (September 1994)

Since the manuscript of this work has been submitted S. Fuchino, S. Koppelberg and S. Shelah have studied σ-filtered Boolean algebras. Their preprint [19] contains among other interesting results the following theorems answering the above questions.

Information 4.2.12 *It it consistent (with the axioms of ZFC) that no infinite complete Boolean algebra is* σ-*filtered.*

Information 4.2.13 *For all σ-filtered Boolean algebras A it holds that*

$$Depth(A) \leq \aleph_1 \quad and \quad |A| \leq |ind\, A|^{\aleph_0}.$$

They also give a very easy proof of 4.2.4.

4.3   The strong Freese/Nation property

In connection with the results of the previous section we introduce the following notion.

Definition 4.3.1 A Boolean algebra A will be said to have the *strong Freese/Nation property*, abbreviated as (SFN), if there is a *directed* family C of pairwise commuting *finite* subalgebras of A such that $A = \bigcup C$. □

It should be obvious that (SFN) implies (FN). At the moment we are not able to either confirm or refute the reverse implication. In the rest of this section we list what we know about this problem.

Proposition 4.3.2 *All projective Boolean algebras enjoy the property (SFN).*

Proof.   Let A be projective. By 1.3.2, there is an open filtration $(A_\alpha)_{\alpha < \lambda}$ of A such that $A_0 = 2$ and each $A_{\alpha+1}$ is generated by $A_\alpha \cup \{a_\alpha\}$ for a suitable element a_α. By induction on α we construct families C_α that witness (SFN) for A_α in such a way that $C_\alpha \subseteq C_\beta$ for all $\alpha < \beta < \lambda$. Then $\bigcup_{\alpha < \lambda} C_\alpha$ is easily seen to witness (SFN) for A.

We can (and must) put $C_0 = \{A_0\}$. At limit steps we take unions (verify that this works!).

The successor step is what really matters. So let C_α be constructed. From $A_\alpha \leq_{rc} A_{\alpha+1}$ it follows that a_α has an upper and a lower projection to A_α, denote them by a_+ and a_-, respectively.

We put
$$C_{\alpha+1} = C_\alpha \cup \{C(a_\alpha) : a_+, a_- \in C \in C_\alpha\}.$$

There should be no problem in checking that $C_{\alpha+1}$ is directed and covers $A_{\alpha+1}$. To see that it is commutative, one has to verify the following three assertions.

(1) $C \downarrow C'$ for $C, C' \in C_\alpha$.

(2) $C \downarrow C'(a_\alpha)$ for $C, C' \in C_\alpha$, $a_+, a_- \in C'$.

(3) $C(a_\alpha) \downarrow C'(a_\alpha)$ for $a_+, a_- \in C, C' \in C_\alpha$.

(1) is trivial, by induction hypothesis, and (2) is similar to (3). So we confine ourselves to the latter.

Consider two typical elements of $C(a_\alpha)$ and $C'(a_\alpha)$, respectively, and assume that they are disjoint, i.e.

$$[(c_1 \wedge a_\alpha) \vee (c_2 - a_\alpha)] \wedge [(c_1' \wedge a_\alpha) \vee (c_2' - a_\alpha)] = 0$$

for $c_1, c_2 \in C$ and $c_1', c_2' \in C'$. Then

$$c_1 \wedge c_1' \wedge a_\alpha = 0 \quad \text{and} \quad c_2 \wedge c_2' - a_\alpha = 0.$$

By minimality of a_+ resp. maximality of a_- we get

$$c_1 \wedge c_1' \wedge a_+ = 0 \quad \text{and} \quad c_2 \wedge c_2' - a_- = 0,$$

hence

$$(c_1 \wedge a_+) \wedge (c_1' \wedge a_+) = 0 \quad \text{and} \quad (c_2 - a_-) \wedge (c_2' - a_-) = 0.$$

As C and C' commute, we find elements $d_1, e_1, d_2, e_2 \in C \cap C'$ such that

$$c_1 \wedge a_+ \leq d_1, \qquad c_1' \wedge a_+ \leq e_1, \qquad d_1 \wedge e_1 = 0,$$

and

$$c_2 - a_- \leq d_2, \qquad c_2' - a_- \leq e_2, \qquad d_2 \wedge e_2 = 0.$$

It follows that

$$[(c_1 \wedge a_\alpha) \vee (c_2 - a_\alpha)] \leq [(d_1 \wedge a_\alpha) \vee (d_2 - a_\alpha)] \in C(a_\alpha) \cap C'(a_\alpha),$$
$$[(c_1' \wedge a_\alpha) \vee (c_2' - a_\alpha)] \leq [(e_1 \wedge a_\alpha) \vee (e_2 - a_\alpha)] \in C(a_\alpha) \cap C'(a_\alpha)$$

and

$$[(d_1 \wedge a_\alpha) \vee (d_2 - a_\alpha)] \wedge [(e_1 \wedge a_\alpha) \vee (e_2 - a_\alpha)] = 0,$$

as desired. $\square$

Knowing this proposition one may be tempted to hope that (SFN) characterizes projectivity. It does not, because of 3.3.9 and the corollary below that will follow at once from

Lemma 4.3.3 *Assume that A and B are commuting subalgebras of some Boolean algebra C. Then $exp\,A$ and $exp\,B$ are also commuting.*

Here $exp\,A$ resp. $exp\,B$ are identified with the subalgebras of $exp\,C$ that consist of all formal sums $c_1 \oplus \ldots \oplus c_n$ whose terms are in A resp. B. From the uniqueness (up to the order of the terms) of the representation of an element of $exp\,C$ as a sum of pairwise distinct non-zero elements of C it follows that

$$exp\,A \cap exp\,B = exp(A \cap B).$$

Before embarking on the commutativity proof, we recall some notation from section 3.3. For $F \subseteq A$ we put

$$U(F) = \{a_1 \oplus \ldots \oplus a_n \in exp\,A : \text{ an odd number of the } a_i \text{ belongs to } F\}.$$

Below we write $U_A(F)$ to stress that we work in the exponential of A. We proved in 3.3.2 that the sets of the form $U(F)$ are exactly the ultrafilters of $exp\,A$ if F runs through the proper (i.e. $0 \notin F$) filters of A. It should be obvious that

$$U_C(F) \cap exp\,A = U_A(F \cap A)$$

for all proper filters F of $C \geq A$.

Now we start with the **proof.** In order to check ($\downarrow 4$), we let V and W be ultrafilters of $exp\,A$ resp. $exp\,B$ such that

$$(*) \qquad V \cap (exp\,A \cap \exp B) = W \cap (exp\,A \cap \exp B).$$

We have to find an extension of $V \cup W$ to an ultrafilter of $exp\,C$.

First we choose proper filters F and G such that $V = U_A(F)$ and $W = U_B(G)$. The desired extension will be found in the form $U_C(H)$, where H is a proper filter of C. It is clear that $U_C(H)$ extends $U_A(F)$ and $U_B(G)$ iff $H \cap A = F$ and $H \cap B = G$. We verify that

$$H = \{c \in C : f \wedge g \leq c \text{ for some } f \in F \text{ and } g \in G\}$$

does the job. It should be clear that H is a filter. To see that it is proper, we need the commutativity of A and B and condition ($*$) which implies

$$F \cap A \cap B = G \cap A \cap B.$$

Assume then that H were not proper. Then $f \wedge g = 0$ for suitable $f \in F \subseteq A$ and $g \in G \subseteq B$. As A and B commute, there are some disjoint $c, d \in A \cap B$ such that $f \leq c$ and $g \leq d$. The first of these relations implies $c \in F \cap A \cap B \subseteq G$, whereas the second relation yields $d \in G$. It follows that $0 = c \wedge d \in G$, which contradicts the properness of G.

By symmetry, it is enough to prove $H \cap A = F$ and, as the reverse inclusion is obvious, even $H \cap A \subseteq F$ will be sufficient. So take an arbitrary $a \in H \cap A$ and let $f \in F, g \in G$ be such that $f \wedge g \leq a$. Then $(f - a) \wedge g = 0$ and, by commutativity,

$$f - a \leq c \quad \text{and} \quad g \leq d \quad \text{for some disjoint } c, d \in A \cap B.$$

From $g \leq d$ we get $d \in G \cap A \cap B \subseteq F$, hence $d \leq -c \leq -(f-a) = -f \vee a \in F$. It follows that $(-f \vee a) \wedge f = a \wedge f \in F$ and, finally $a \wedge f \leq a \in F$, as claimed. $\square$

Corollary 4.3.4 *The properties (WFN) and (SFN) are preserved by exponentials.*

Notice that for (SFN) it is essential that the commuting family in A is directed, otherwise we would not get $exp\,A = \bigcup \{exp\,C : C \in \mathcal{C}\}$.

We end this section by showing that our second concrete example, namely 3.4.7, of a non-projective rc-filtered Boolean algebra also has the property (SFN). This will be an immediate consequence of

Lemma 4.3.5 *Assume that A and B are commuting subalgebras of some Boolean algebra C. Then the symmetric powers $SP^2(A)$ and $SP^2(B)$ are commuting subalgebras of $SP^2(C)$.*

Recall from section 3.4 that $SP^2(C)$ is the subalgebra of $C \otimes C$ consisting of all symmetric elements, i.e. those that are fixed by the automorphism $\sigma = \sigma_C$, which maps $c_1 \otimes c_2$ to $c_2 \otimes c_1$. If $A \leq C$, then $A \otimes A$ is naturally embedded into $C \otimes C$ and $\sigma_A = \sigma_C \restriction A \otimes A$. It follows that $SP^2(A)$ is naturally isomorphic to $SP^2(C) \cap A \otimes A$. And in that sense the lemma is to be understood.

Proof. As a first step we show

CLAIM　*The subalgebras $A \otimes A$ and $B \otimes B$ of $C \otimes C$ commute.*

Indeed, take typical elements $\bigvee_{i=1}^m (a_i^1 \otimes a_i^2)$ of $A \otimes A$ and $\bigvee_{j=1}^n (b_j^1 \otimes b_j^2)$ of $B \otimes B$, respectively. Assuming that they are disjoint, we get

$(*)$　for all i,j　$a_i^1 \wedge b_j^1 = 0$　or　$a_i^2 \wedge b_j^2 = 0$.

As A and B commute, we can find elements $c_{ij}^k, d_{ij}^k \in A \cap B$ such that for all i,j,k　$a_i^k \leq c_{ij}^k$ and $b_j^k \leq d_{ij}^k$, and

$(**)$　if $a_i^k \wedge b_j^k = 0$, then $c_{ij}^k \wedge d_{ij}^k = 0$.

It follows that

$$\bigvee_{i=1}^m (a_i^1 \otimes a_i^2) \quad \leq \quad \bigvee_{i=1}^m [(\textstyle\bigwedge_{j=1}^m c_{ij}^1) \otimes (\bigwedge_{j=1}^m c_{ij}^2)] \quad \in \quad (A \otimes A) \cap (B \otimes B)$$
$$\text{and}$$
$$\bigvee_{j=1}^n (b_j^1 \otimes b_j^2) \quad \leq \quad \bigvee_{j=1}^n [(\textstyle\bigwedge_{i=1}^m d_{ij}^1) \otimes (\bigwedge_{i=1}^m d_{ij}^2)] \quad \in \quad (A \otimes A) \cap (B \otimes B)$$

Moreover,

$$\bigvee_{i=1}^m [(\textstyle\bigwedge_{j=1}^m c_{ij}^1) \otimes (\bigwedge_{j=1}^m c_{ij}^2)] \wedge \bigvee_{j=1}^n [(\textstyle\bigwedge_{i=1}^m d_{ij}^1) \otimes (\bigwedge_{i=1}^m d_{ij}^2)]$$
$$\leq \bigvee_{i=1}^m \bigvee_{j=1}^n [(c_{ij}^1 \wedge d_{ij}^1) \otimes (c_{ij}^2 \wedge d_{ij}^2)] = 0,$$

by $(*)$ and $(**)$. The claim is proved.

Now we prove the result proper. Let $p \in SP^2(A)$ and $q \in SP^2(B)$ be given such that $p \leq q$. The claim yields $x \in (A \otimes A) \cap (B \otimes B)$ such that $p \leq x \leq q$. It follows that $p = \sigma(p) \leq \sigma(x) \leq \sigma(q) = q$, hence $p \leq x \vee \sigma(x) \leq q$. Now, $x \vee \sigma(x)$ is symmetric and belongs to $(A \otimes A) \cap (B \otimes B)$, i.e. to $SP^2(A) \cap SP^2(B)$, as desired. $\square$

As above we conclude that

Corollary 4.3.6 *The properties (WFN) and (SFN) are preserved by symmetric powers.*

Chapter 5

Weakly projective and regularly filtered algebras

In this chapter we study two more adequate pairs. We shall be mainly concerned with what we call[1] weakly projective Boolean algebras, i.e. those Boolean algebras that are co-complete with a projective Boolean algebra. The material of this chapter is of some interest in connection with the notion of forcing[2]. Indeed, forcing with either of two co-complete Boolean algebras yields exactly the same generic extensions (of models of set theory). Moreover, one interpretation of theorem 5.2.2 says that forcing with some weakly projective Boolean algebra is always equivalent to some Cohen forcing.

The results of this chapter are mostly due to the second author. In [36] Koppelberg treats part of the material in the Boolean algebraic language. There she introduced the phrase 'Cohen algebra' just for those Boolean algebras who are co-complete to products of free algebras.

So, our our weakly projective Boolean algebras turn out to be Koppelberg's ccc Cohen algebras.

The last section of the chapter deals with the class of Boolean algebras that is adequate for the class of regular embeddings.

5.1   Completions and co-completeness

The reader is supposed to know that every Boolean algebra can be densely embedded into a complete Boolean algebra. Moreover, this embedding is unique in the following sense.

If $A \leq_d B$ and $A \leq_d C$, where B and C are complete, then there exists an isomorphism $\varphi : B \to C$ whose restriction to A is the identity.

---

[1] Again in want of a better name.
[2] To assure the non-specialist: this will play no explicit role below.

This fact gives us the right to speak about *the* completion of a Boolean algebra A and to denote it by $\overline{A}$. One way of obtaining $\overline{A}$ is to represent A as $Clop(X)$ for a compact and zero-dimensional space X and to take $\overline{A} = RO(X)$, the Boolean algebra of regular open subsets of X. For more details and a proof of the above fact the reader may consult subsection 4.3 of [34].

Also without proof we quote

Sikorski's Extension Theorem (5.9 in [34]) *Assume $A \leq B$ and let C be complete. Then every homomorphism $A \to C$ can be extended to a homomorphism $B \to C$.*

If $\varphi : A \to B$ is a homomorphism, we can first conceive it as a mapping into $\overline{B}$ and then use Sikorski's Theorem to extend it to a homomorphism $\overline{\varphi} : \overline{A} \to \overline{B}$. We call $\overline{\varphi}$ a *lifting* of φ. Notice the indefinite article, φ can have many liftings, because the extension provided by Sikorski's Theorem is usually not unique. Fortunately, any lifting of an injective homomorphism is injective again. Indeed, if $\overline{\varphi}(\bar{a}) = 0$ for some non-zero $\bar{a} \in \overline{A}$, then taking $a \in A$ such that $0 < a \leq \bar{a}$ we get $\varphi(a) = \overline{\varphi}(a) \leq \overline{\varphi}(\bar{a}) = 0$, contradicting the injectivity of φ.

In the following important case the lifting is unique.

Lemma 5.1.1 *If $A \leq_{reg} B$, then the inclusion map $\iota : A \to B$ sending $a \in A$ to itself has a unique lifting $\bar{\iota} : \overline{A} \to \overline{B}$ given by the formula*

$$\bar{\iota}(\bar{a}) = \bigvee^{\overline{B}} \{a \in A : a \leq \bar{a}\}.$$

Moreover, $\bar{\iota}$ turns out to be a homomorphism also with respect to the infinitary operations.

Proof. Let $\bar{\iota} : \overline{A} \to \overline{B}$ be any lifting of ι. By the above remark, $\bar{\iota}$ is injective.

CLAIM $\bar{\iota}(\overline{A}) \leq_{reg} \overline{B}$

To see this, consider any non-zero $\bar{b} \in \overline{B}$. Choose $b \in B$ such that $0 < b \leq \bar{b}$ and then a non-zero $a \in A$ such that $A \restriction (a - b) = \{0\}$. These elements exist because $B \leq_d \overline{B}$ and $A \leq_{reg} B$.

To verify $A \restriction (a - b) = \{0\}$, assume that there were some $\bar{c} \in \overline{A}$ such that $0 < \bar{c} \leq a$ and $\bar{c} \wedge \bar{b} = 0$. By density, we can find $c \in A$ such that $0 < c \leq \bar{c}$. But then $c \in A \restriction (a - b) = \{0\}$, which is the desired contradiction.

By 1.2.16, the claim yields that $\bar{\iota}$ preserves infinite joins. Indeed, for an arbitrary subset M of $\overline{A}$ we have $\bar{\iota}(\bigvee^{\overline{A}} M) = \bigvee^{\bar{\iota}(\overline{A})} \bar{\iota}(M)$, because $\bar{\iota}$ is an order isomorphism, and $\bigvee^{\bar{\iota}(\overline{A})} \bar{\iota}(M) = \bigvee^{\overline{B}} \bar{\iota}(M)$, by regularity. The formula $\bigwedge M = - \bigvee -M$ shows that infinite meets are also preserved. Now, for each $\bar{a} \in \overline{A}$,

$$\bar{\iota}(\bar{a}) = \bar{\iota}(\bigvee^{\overline{A}} \{a \in A : a \leq \bar{a}\}) = \bigvee^{\overline{B}} \{\bar{\iota}(a) : a \in A;\ a \leq \bar{a}\} = \bigvee^{\overline{B}} \{a \in A : a \leq \bar{a}\}.$$

From this formula, uniqueness follows at once. □

The lemma will allow us to associate with every regular embedding $A \leq_{reg} B$ the diagram

$$A \quad \leq \quad B$$
$$\wedge|_d \qquad \wedge|_d$$
$$\overline{A} \quad \leq \quad \overline{B}.$$

Moreover, with every regular filtration $(A_\alpha)_{\alpha < \lambda}$ of some Boolean algebra A we can consider $(\overline{A_\alpha})_{\alpha < \lambda}$ as a chain of subalgebras of $\overline{A}$. Notice however, that this chain is not a filtration of $\overline{A}$, for $\bigcup_{\alpha < \lambda} \overline{A_\alpha}$ need not equal $\overline{A}$, but is only a dense subalgebra. For the same reason the chain of completions will not be continuous.

After all these preparations we come to the central notion of this chapter.

Definition 5.1.2 Two Boolean algebras are said to be *co-complete* (in symbols $A \sim B$) if they have isomorphic completions. □

The uniqueness of completions immediately yields reflexivity, symmetry and transitivity. So co-completeness is an equivalence relation.

Observation 5.1.3 *If $A \leq_d B$, then $A \sim B$.*

Indeed, by transitivity of dense embeddings, $A \leq_d B \leq_d \overline{B}$ yields $A \leq_d \overline{B}$, hence $\overline{A} \simeq \overline{B}$.

Corollary 5.1.4 *The Boolean algebras A and B are co-complete if and only if there exist A', B', and C such that $A \simeq A' \leq_d C \geq_d B' \simeq B$.*

Proof. If $A \sim B$, then either $\overline{A}$ or $\overline{B}$ can be taken for C. The other way round: $A \simeq A' \leq_d C \geq_d B' \simeq B$ implies $\overline{A} \simeq \overline{A'} \simeq \overline{C} \simeq \overline{B'} \simeq \overline{B}$. □

As an exercise the reader may prove that B is co-complete to a given atomic Boolean algebra A if and only if B is also atomic and has the same number of atoms as A. This shows that co-completeness is a rather coarse relation. Nevertheless there are some properties that co-complete algebras must share.

Lemma 5.1.5 *Assume $A \sim B$.*

(1) $c(A) = c(B)$. *So, in particular, the ccc is preserved.*

(2) *A and B have the same calibers.*

(3) $\pi(A) = \pi(B)$.

(4) *If $\pi(A \restriction a) = \pi(A)$ for all non-zero $a \in A$, then $\pi(B \restriction b) = \pi(B)$ for all non-zero $b \in B$.*

The property expressed in (4) will be called π-*homogeneity* below.

Proof. Recalling the definitions of cellularity, caliber, and π-weight (cf. section 2.6), all three assertions are easy to prove. Just to give one example, we do (3). By 5.1.4, we can assume $A \leq_d C \geq_d B$. Let X be a dense subset of $A \setminus \{0\}$ of cardinality $\pi(A)$. From $X \subseteq C \geq_d B$ it follows that for each $x \in X$ there exists some $f(x) \in B$ such that $0 < f(x) \leq x$. Then $Y = f(X)$ is dense in B. Indeed, given $b \in B \setminus \{0\}$ we first choose $a \in A \setminus \{0\}$ and then $x \in X$ such that $x \leq a \leq b$. Clearly, $Y \ni f(x) \leq b$.

Moreover $|Y| \leq |X| = \pi(A)$, hence $\pi(B) \leq \pi(A)$. The reverse inequality follows by symmetry. $\square$

Recall that the weak product of a family of Boolean algebras has been introduced in 2.5.2.

Lemma 5.1.6 *Let $(a_i)_{i \in I}$ be a family of pairwise disjoint non-zero elements of some Boolean algebra A. If each non-zero $a \in A$ intersects some a_i, then A is co-complete to the weak product $\prod_{i \in I}^{w}(A \restriction a_i)$.*

Proof. The assumption yields that $\bigcup_{i \in I}(A \restriction a_i)$ generates a dense subalgebra of A. Using the disjointness of the family, it is easy to verify that $a \mapsto (a \wedge a_i)_{i \in I}$ is an isomorphism

$$\langle \textstyle\bigcup_{i \in I}(A \restriction a_i)\rangle \longrightarrow \prod_{i \in I}^{w}(A \restriction a_i).$$

The assertion now follows from 5.1.3. $\square$

Using the easy fact that if $A_i \leq_d C_i$ for all $i \in I$, then $\prod_{i \in I}^{w} A_i \leq_d \prod_{i \in I}^{w} C_i$, we get

Corollary 5.1.7 *Let $(a_i)_{i \in I}$ and $(b_i)_{i \in I}$ be two families of non-zero elements of Boolean algebras A and B, respectively. If $A \restriction a_i \sim B \restriction b_i$ for all $i \in I$ and both families satisfy the conditions of the lemma above, then $A \sim B$.*

The notion of co-completeness can be extended to embeddings. There are several equivalent ways to do this. We adopt the following

Definition 5.1.8 The embeddings $A_1 \leq B_1$ and $A_2 \leq B_2$ will be called co-complete if there is a third embedding $C \leq D$ and injective homomorphisms $B_1 \xrightarrow{\beta_1} D \xleftarrow{\beta_2} B_2$ such that the images $\beta_i(B_i)$ are dense in D and the images $\beta_i(A_i)$ are dense in C. $\square$

The following diagram illustrates this definition.

$$\begin{array}{ccccccccc}
B_1 & \xrightarrow{\beta_1} & B_1' & \leq_d & D & \geq_d & B_2' & \xleftarrow{\beta_2} & B_2 \\
\vee| & & & & \vee| & & & & \vee| \\
A_1 & \xrightarrow{\beta_1} & A_1' & \leq_d & C & \geq_d & A_2' & \xleftarrow{\beta_2} & A_2
\end{array}$$

Whenever convenient we can assume that either β_1 or β_2 is the identity.

Among the few things that co-complete embeddings must have in common is regularity.

Observation 5.1.9 *If $A_1 \leq B_1$ and $A_2 \leq B_2$ are co-complete and $A_1 \leq_{reg} B_1$, then $A_2 \leq_{reg} B_2$.*

Indeed, as regularity does not change under isomorphisms, we can assume the following situation:

$$B_1 \qquad \leq_d \quad D \quad \geq_d \quad B_2$$
$$\mathsf{VI}^{reg} \qquad\qquad \mathsf{VI} \qquad\quad \mathsf{VI}$$
$$A_1 \qquad \leq_d \quad C \quad \geq_d \quad A_2.$$

The result then follows by a simple diagram chase. If a non-zero $b_2 \in B_2$ is given, then one chooses in this order non-zero $b_1 \in B_1'$, $a_1 \in A_1'$, and $a_2 \in A_2$ such that $b_1 \leq b_2$, $A_1' \restriction (a_1 - b_1) = \{0\}$, and $a_2 \leq a_1$. It is then routine to check $A_2 \restriction (a_2 - b_2) = \{0\}$. $\square$

For regular embeddings there is a more concise and natural description of co-completeness.

Observation 5.1.10 *Two regular embeddings $A_1 \leq_{reg} B_1$ and $A_2 \leq_{reg} B_2$ are co-complete if the associated embeddings of their completions $\overline{A_1} \leq \overline{B_1}$ and $\overline{A_2} \leq \overline{B_2}$ are isomorphic, i.e. if there is an isomorphism $\varphi : \overline{B_1} \to \overline{B_2}$ which maps $\overline{A_1}$ onto $\overline{A_2}$.*

All co-complete pairs of embeddings that occur below will be regular and we take advantage of the above remark, leaving its proof, however, to the reader.

Topological duality. The topologically dual notion to completion is that of an absolute or Gleason space. Our co-completeness corresponds to what, following Ponomarev (cf. [45]), topologists call co-absoluteness.

5.2 Weakly projective algebras are Cohen algebras

Definition 5.2.1 We say that a Boolean algebra is *weakly projective* iff it is co-complete with a projective Boolean algebra. $\square$

The main result of this section determines weakly projective Boolean algebras up to co-completeness. In the formulation below we consider two-element Boolean algebra as free. In fact, this is in accordance with the usual definition — they are free over the empty set of generators.

Theorem 5.2.2 *A Boolean algebra is weakly projective iff it is co-complete with a product of at most countably many free Boolean algebras.*

Notice that in the above formulation products can be replaced by weak products because $\Pi^w_{i \in I} A_i$ is always dense in $\Pi_{i \in I} A_i$. As countable weak products of free Boolean algebras are projective, this observation settles one direction of the theorem.

The same remark reduces the other direction to the assertion that every weakly projective Boolean algebra is co-complete with a weak product of at most countably many free Boolean algebras. In fact, we need not bother about countability for, if A is weakly projective, it has the ccc, by 5.1.5, hence $A \sim \Pi^w_{i \in I} F_i$ implies

$$|I| \leq c(\Pi^w_{i \in I} F_i) = c(A) = \aleph_0.$$

The main work will be with the following special case.

Proposition 5.2.3 *Let A be a weakly projective algebra of infinite π-weight κ. Then $A \sim \mathrm{Fr}\,\kappa$ iff A is π-homogeneous.*

Before proving the proposition we derive the theorem from it. Let A be the given weakly projective algebra. We construct a weak product $\Pi^w_{k \in K} F_k$ of free Boolean algebras, which is coabsolute with A and has the additional properties

(1) $|\{k \in K : F_k = 2\}|$ is the number of atoms of A,   and

(2) if an infinite free algebra occurs as F_k, then it occurs infinitely often.

We shall say that $a \in A$ is π-homogeneous if either a is an atom or $A\!\restriction a$ is atomless and π-homogeneous.

The set H of all π-homogeneous elements is dense in A. Indeed, let $0 < b \in A$ be given. If there is an atom below b, this atom belongs to H. If there is no atom, choose $0 < a \leq b$ in such a way that the cardinal $\pi(A\!\restriction a)$ becomes minimal. Then, clearly, $a \in H$.

Now we take a maximal disjoint subset K of H. By density of H and maximality of K, every non-zero element of A intersects some $k \in K$.

From 5.1.6 we get $A \sim \Pi^w_{k \in K}(A\!\restriction k)$. If k is an atom, then $A\!\restriction k = F_k = 2$ is free over $\emptyset$. If k is not an atom, then $A\!\restriction k \sim F_k$ for an infinite free algebra, by 5.2.3. Finally, $A \sim \Pi^w_{k \in K} F_k$, by 5.1.7.

Extra condition (1) will be satisfied automatically, because all atoms must belong to K. Condition (2), however, may not be satisfied by our first choice of K. To remedy this, we replace each non-atom $k \in K$ by a an infinite maximal disjoint subset of the atomless algebra $A\!\restriction k$. After that the new K remains maximal disjoint in A and still consists of π-homogeneous elements. But now every infinite cardinal that occurs as $\pi(A\!\restriction k)$ occurs infinitely often. $\square$

To benefit from the extra work in the preceeding proof, we introduce the following

Definition 5.2.4 The π-*spectrum* of the Boolean algebra A is the set

$$\{\pi(A\!\restriction a) : A\!\restriction a \text{ is infinite and } \pi\text{-homogeneous }\}. \quad \square$$

Notice that A has empty π-spectrum iff it is atomic. Conditions (1) and (2) in the above proof show that the weak product constructed there depends only on the number of atoms and the π-spectrum of A. As co-completeness is a transitive relation, this observation yields

Corollary 5.2.5 *Two weakly projective algebras are co-complete if they have the same number of atoms and equal π-spectra.*

As an exercise in the style of the proof of 5.1.5 the reader may prove that both conditions are also necessary for co-completeness, even if the algebras are not weakly projective.

The whole rest of this section will be occupied with the proof of 5.2.3. The necessity is immediate. As $\mathrm{Fr}\,\kappa$ is π-homogeneous, every $A \sim \mathrm{Fr}\,\kappa$, must be π-homogeneous, by 5.1.5.

Assume now that A is weakly projective, π-homogeneous and has π-weight $\kappa \geq \aleph_0$. Replacing A if necessary by a co-complete projective algebra, we can assume that A itself is projective. By 5.1.5, this does not affect neither the π-weight nor the π-homogeneity. By 2.6.1, for projective Boolean algebras π-weight equals cardinality, so $|A{\restriction}\,a| = |A| = \kappa$ for all non-zero $a \in A$.

In particular, A is atomless. This alone settles the countable case.

If $\kappa = |A|$ is regular uncountable, the proposition can easly be deduced from earlier results. We give this proof for the convenience of those readers whose ambitions don't reach $\aleph_\omega$. It starts with the following

CLAIM   *There is a family $(a_i)_{i \in I}$ of pairwise disjoint elements of A such that*

    (1) *$A{\restriction}\, a_i \simeq \mathrm{Fr}\kappa$ for each i,*

    (2) *each non-zero $a \in A$ intersects some a_i, and*

    (3) *the index set I is countably infinite.*

To see this, let P be any prime ideal of A. Choose a maximal family $(a_i)_{i \in I}$ of pairwise disjoint elements of P such that $A{\restriction}\, a_i \simeq \mathrm{Fr}\,\kappa$. Then condition (1) is automatic. To prove (2) we consider an arbitrary $a \in A$. As A is atomless, $a' \in P$ for some non-zero $a' \leq a$. By 2.4.1, $A{\restriction}\, a'$ is projective. Moreover, $|A{\restriction}\, a'| = \kappa$. Together with the regularity of κ, 2.7.8 yields $a'' \leq a'$ such that $a'' \simeq \mathrm{Fr}\,\kappa$. By maximality, a'' intersects some a_i.

As A satisfies the ccc, I is countable. If I were finite, then (2) would yield $1 = \bigvee_{i \in I} a_i \in P$, which contradicts the choice of P as a prime ideal. Condition (3) is also proved.

Together with corollary 5.1.7 the claim yields $A \sim \mathrm{Fr}\,\kappa$. Indeed, $\mathrm{Fr}\,\kappa$ satisfies the same assumptions as A. Therefore, an analogous family exits in $\mathrm{Fr}\,\kappa$.

Unfortunately, the above argument does not work for singular κ. That is why we now give another proof, which works for all uncountable κ. It has the additional advantage of not using results from 2.7, which makes it, on the whole, not more complicated than the one above. We follow the same general pattern as the proof of Ščepin's Theorem 1.4.11, which is not surprising given that the statements are similar.

We shall use extensions that are co-complete with free ones. The following observation will be useful.

Observation 5.2.6 *The embedding $A \leq B$ is co-complete with a free extension $A' \leq_{free} B'$ iff there exists a free subalgebra $F \leq \overline{B}$, which is independent of A and such that $\langle A \cup F \rangle \leq_d \overline{B}$.*

Indeed, one direction is obvious. If F exists, then the diagram

$$
\begin{array}{ccccc}
B & \leq_d & \overline{B} & \geq_d & \langle A \cup F \rangle \\
\text{\Large V|} & & \text{\Large V|} & & \text{\Large V|}^{free} \\
A & = & A & = & A
\end{array}
$$

shows that $A \leq_{free} \langle A \cup F \rangle$ is co-complete with $A \leq B$.

Assume then that $A \leq B$ and $A' \leq_{free} B'$ are co-absolute. Replacing A' and B' by isomorphic copies if necessary, we can assume

$$
\begin{array}{ccc}
\overline{A} & \leq & \overline{B} \\
\text{\Large V|}^d & & \text{\Large V|}^d \\
A' & \leq & B' = \langle A' \cup F \rangle,
\end{array}
$$

where $F \leq B'$ is a free subalgebra witnessing $A' \leq_{free} B'$.

We check that F is as desired. To see that it is independent of A, let $0 < a \in A$ and $0 < f \in F$ be given. As A' is dense in $\overline{A} \supseteq A$, we find $a' \in A'$ such that $0 < a' \leq a$. As A' and F are independent, $0 < a' \wedge f \leq a \wedge f$, as desired.

To see that $\langle A \cup F \rangle \leq_d \overline{B}$, let a non-zero $b \in \overline{B}$ be given. As $\langle A' \cup F \rangle$ is dense in $\overline{B}$, we find $0 < a' \wedge f \leq b$. The density of A in $\overline{A} \supseteq A'$ then allows us to pick $a \in A$ such that $0 < a \leq a'$. Clearly,

$$
\langle A \cup F \rangle \ni a \wedge f \leq a' \wedge f \leq b.
$$

Moreover, $a \wedge f \neq 0$, by the already proved independence of A and F. $\square$

The following result is analogous to 1.4.4.

Lemma 5.2.7 *Let $(A_\alpha)_{\alpha < \lambda}$ be a well-ordered filtration of A. If A_0 is co-complete with a free Boolean algebra and each $A_\alpha \leq A_{\alpha+1}$ is co-complete with a free extension, then A is co-complete with a free Boolean algebra.*

Proof. We put $A_\lambda = A$ and construct a sequence $(F_\alpha)_{\alpha \leq \lambda}$ of free algebras such that $F_\alpha \leq_d \overline{A}_\alpha$ and, to keep the induction going, $F_\alpha \leq_{free} F_{\alpha+1}$. The existence of F_λ then proves the lemma.

$F_0 \leq_d \overline{A}_0$ exists by assumption. Assume that F_α is constructed and look at the diagram (here we use lemma 5.1.1).

$$
\begin{array}{ccc}
A_\alpha & \leq & A_{\alpha+1} \\
\text{\Large $\wedge$|}^d & & \text{\Large $\wedge$|}^d \\
\overline{A}_\alpha & \leq & \overline{A}_{\alpha+1} \\
\text{\Large V|}^d & & \\
F_\alpha & &
\end{array}
$$

As $A_\alpha \leq A_{\alpha+1}$ is co-complete with a free extension, the above observation allows us to take $F \leq \overline{A}_{\alpha+1}$, independent of A_α and such that $\langle A_\alpha \cup F \rangle$ is dense in $\overline{A}_{\alpha+1}$. An easy verification then shows that $F_{\alpha+1} = \langle F_\alpha \cup F \rangle$ is as desired.

Assume now that β is a limit ordinal and that F_α have been constructed for $\alpha < \beta$. We put $F_\beta = \bigcup_{\alpha<\beta} F_\alpha$. Then F_β is free, by 1.4.4. Moreover, by construction, F_β is dense in $\bigcup_{\alpha<\beta} \overline{A}_\alpha$, which is dense in $\bar{A}_\beta$. □

The role of Sirota's lemma 1.4.10 will be played by the following result, which was first proved by Bandlow [3] under the additional assumption that A satisfies the ccc. As it is interesting in its own right, we qualify it as

Theorem 5.2.8 *An arbitrary embedding $A \leq B$ of countable π-weight is co-complete with a free extension iff it is regular and*

(∗) *for all non-zero $b \in B$ there are non-zero disjoint $b_0, b_1 \leq b$ such that $A{\restriction} - b_1 = A{\restriction} - b_2$.*

Proof. Given that the necessity of the condition will play no further role, we leave its easy verification to the reader.

Assume then $A \leq_{reg} B$ and (∗). We shall work in the completion $\overline{B}$ of B. All infinite joins that occur below are taken there. We use $\dot\vee$ to denote disjoint union. First we strengthen (∗) by establishing that

(∗∗) *all non-zero $b \in \overline{B}$ can be split into $b = b_0 \dot\vee b_1$ in such a way that $A{\restriction} - b_1 = A{\restriction} - b_2 = A{\restriction} - b$.*

To see this, let $0 < b \in \overline{B}$ be given. By Zorn's Lemma, there exists a maximal with respect to inclusion collection C of pairs such that

$$0 < c_0, c_1 \leq b, \quad c_0 \wedge c_1 = 0, \quad A{\restriction} - c_0 = A{\restriction} - c_1, \text{ and } (c_0 \vee c_1) \wedge (d_0 \vee d_1) = 0$$

for all $(c_0, c_1), (d_0, d_1) \in C$. The reader will then easily verify that by putting

$$b_0 = \bigvee \{c_0 : (c_0, c_1) \in C\} \quad \text{and} \quad b_1 = \bigvee \{c_1 : (c_0, c_1) \in C\}$$

we arrive at the desired splitting of b. Condition (∗) is needed to show $b = b_1 \vee b_2$.

Now we use (∗∗) to prove the main technical device.

CLAIM   *Assume that $u, v \in \overline{B}$ are such that $u \wedge v \neq 0$. Then u can be written as $u = u_0 \dot\vee u_1$ in such a way that*

(1) $A{\restriction} - u_0 = A{\restriction} - u_1 = A{\restriction} - u$ *and*

(2) *for all $a \in A$ such that $a \wedge u \wedge v \neq 0$ there exists $a' \in A$ such that $0 < a' \wedge u_0 \leq a \wedge v$.*

To construct u_0 and u_1 one first splits u into the tree parts

$$r = \bigvee\{x \wedge u : x \in A \text{ and } x \wedge u \leq -v\}, \quad h = \bigvee\{x \wedge u : x \in A \text{ and } x \wedge u \leq v\},$$

and $g = u - (r \vee h)$. Clearly, $h \leq u \wedge v$, $r \leq u - v$ and $u = r \dot{\vee} g \dot{\vee} h$.

Using (∗∗), we further split $r = r_0 \dot{\vee} r_1$ and $h = h_0 \dot{\vee} h_1$ in such a way that

$$A\!\upharpoonright - r_0 = A\!\upharpoonright - r_1 = A\!\upharpoonright - r \quad \text{and} \quad A\!\upharpoonright - h_0 = A\!\upharpoonright - h_1 = A\!\upharpoonright - h.$$

The following picture may be helpful to visualize what happens. (u is the rectangle, v the triangle, and A is the algebra consisting of vertical stripes.)

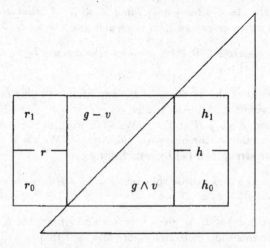

The souhgt-for elements can now be explicitly written down:

$$u_0 = r_0 \dot{\vee} (g \wedge v) \dot{\vee} h_0 \quad \text{and} \quad u_1 = r_1 \dot{\vee} (g - v) \dot{\vee} h_1.$$

It should be clear that they split u. To verify condition (1), we can confine ourselves to $A\!\upharpoonright - u_0 = A\!\upharpoonright - u$, by symmetry. From $u_0 \leq u$ we immediately get $A\!\upharpoonright - u \subseteq A\!\upharpoonright - u_0$. To check the reverse inclusion, consider $a \in A\!\upharpoonright - u_0$, i.e. $a \wedge u_0 = 0$. Then $a \wedge h_0 = 0$ and $a \wedge r_0 = 0$, hence $a \wedge h = 0$ and $a \wedge r = 0$.

Substituting $g \wedge v = (u \wedge v) - [(h \vee r) \wedge v] = (u \wedge v) - h$ into $a \wedge (g \wedge v) = 0$ and using $a \wedge h = 0$ yields

$$a \wedge [(u \wedge v) - h] = (a \wedge u \wedge v - a \wedge h) = a \wedge u \wedge v = 0, \quad \text{i.e. } a \wedge u \leq -v.$$

So, $a \wedge u$ is a joinand of r. Consequently, $a \wedge u \leq r$. On the other hand, $a \wedge r = 0$. It follows that $a \wedge u = 0$, in other words $a \in A\!\upharpoonright - u$.

Now we prove (2). Assume $a \wedge u \wedge v \neq 0$ and use $A \leq_{reg} \overline{B}$ to pick a non-zero $c \in A$ such that

$$A\!\upharpoonright (c - (a \wedge u \wedge v)) = \{0\}.$$

We show that $a' = c \wedge a$ is as required in (2).

Clearly $a' > 0$, for, otherwise $c \in A\!\restriction (c - (a \wedge u \wedge v)) = \{0\}$. Next $a' \wedge u_0 > 0$. Because from $a' \wedge u_0 = 0$ we would get $a' \wedge u = 0$, by (1), hence

$$a' \in A\!\restriction (c - a \wedge u \wedge v) = \{0\},$$

a contradiction. To see that $a' \wedge u_0 \le a \wedge v$, consider any $x \in A$ such that $x \wedge u \le -v$. Then $a' \wedge x \in A\!\restriction (c - (a \wedge u \wedge v)) = \{0\}$ and, therefore, $a' \wedge x \wedge u \le a' \wedge x = 0$. This being true for all $x \wedge u$, we get $a' \wedge r = 0$ and, consequently,

$$a' \wedge u_0 = a' \wedge (u_0 - r) \le a' \wedge u_0 \wedge v \le a \wedge v,$$

as desired. The claim is proved.

After these technical preparations we start the construction of the required free algebra $F \le \overline{B}$. Using $\pi(B/A) \le \aleph_0$ we fix a countable subalgebra V of B such that $\langle A \cup V \rangle \le_d B \le_d \overline{B}$. Let $\{v_n : n < \omega\}$ enumerate the non-zero elements of V in such a way that each v occurs infinitely often as v_n.

Now by induction on the length of the finite $0 - 1$-sequence $\eta \in 2^{<\omega}$ we construct non-zero elements u_η of $\overline{B}$ such that

(o) $u_\emptyset = 1$,

(i) $u_\eta = u_{\eta 0} \,\dot{\vee}\, u_{\eta 1}$,

(ii) $A\!\restriction - u_\eta = \{0\}$.

Moreover, in the construction we take care that

(iii) for all $a \in A$ and $n < \omega$ such that $a \wedge v_n > 0$ there are $a' \in A$ and $\eta \in 2^{<\omega}$ such that $0 < a' \wedge u_\eta \le a \wedge v_n$.

Parallel to the u_η we construct an auxiliary sequence n_η of natural numbers.

Condition (o) leaves no choice for the start: $u_\emptyset = 1$.

Assume then that u_η and n_μ are found for some η and all proper initial segments μ of η. From (ii) we have $u_\eta \ne 0$. As $\langle A \cup V \rangle$ is dense in $\overline{B}$, there are $v \in V$ which intersect u_η. Given that each v occurs infinitely often as v_n, it makes sense to define

$$n_\eta = \min\{n : v_n \wedge u_\eta > 0 \text{ and } n \notin \{n_\mu : \mu \text{ an initial segment of } \eta\}\}.$$

It follows that $u_\eta \wedge v_{n_\eta} > 0$ and the claim provides us with a splitting $u_\eta = u_{\eta 0} \,\dot{\vee}\, u_{\eta 1}$. The construction is finished.

The claim guarantees conditions (i) and (ii) at each stage. To see that the (rather global) condition (iii) is also satisfied, we assume $a \wedge v_n \ne 0$.

From (o) and (i) it follows that $\bigvee \{u_\eta : \text{length of } \eta = n\} = 1$, for all n. This allows us to pick, by induction, sequences $\emptyset, \eta_1, \eta_2, \ldots$, where each η_{i+1} extends η_i by one term and each u_{η_i} intersects $a \wedge v_n$. Then n is an eligible number if it comes to the determination of $n_\emptyset, n_{\eta_1}, n_{\eta_2}, \ldots$. If it is not chosen, then because there are still smaller eligible numbers. But, after n steps at the latest, this reason has expired and n becomes n_{η_i}.

As $u_{\eta,0}$ was chosen in accordance with the claim, we get $a' \in A$ such that $0 < a' \wedge u_{\eta,0} \leq a \wedge v_{n_{\eta_i}}$, as demanded in (iii).

We are left with collecting the results. Let F denote the subalgebra of $\overline{B}$ generated by $\{u_\eta : \eta \in 2^{<\omega}\}$. A moments reflection shows that the non-zero elements of F are finite unions of u_η's. So F is atomless and, being countable, also free. Condition (ii) implies that F is independent of A and condition (iii) says that $(A \cup F)$ is dense in $\overline{B}$. So, by 5.2.6, we are done. $\square$

We still need a

Lemma 5.2.9 Assume $T \leq_{rc} A$ and $|T| < |A\restriction a|$ for all non-zero $a \in A$. Then there is a dense subset M of T such that $A\restriction m$ contains an element which is independent of $T\restriction m$, for all $m \in M$.

Proof[3]. Let $0 < t \in T$ be given. The assumption yields some $a \in (A\restriction t) \setminus T$. We put $m = q(a) - p(a)$, where q and p denote the upper and lower projections associated with $T \leq_{rc} A$. From $a \leq t$ we get $m \leq q(a) \leq t$. Moreover, $m > 0$, for, otherwise, $p(a) \leq a \leq q(a)$ would yield $p(a) = a = q(a) \in T$, a contradiction.

It will be sufficient to show that $c = m \wedge a \in A\restriction m$ is independent with $T\restriction m$. Let $0 < s \in T\restriction m$ be given. Then $s \leq m \leq q(a)$ and

$$0 = s \wedge c = s \wedge (m \wedge a) = s \wedge a$$

would yield $a \leq q(a) - s < q(a)$, which contradicts the minimality of $q(a)$. Symmetrically, $s \leq m \leq -p(a)$ shows that

$$0 = s - c = s - ((m \wedge a) = s - a, \text{ i.e. } s \leq a$$

would yield $p(a) < p(a) \vee s \leq a$, contradicting the maximality of $p(a)$ below a. $\square$

Now we are ready to **prove** the general case of 5.2.3. As we did in the special case (for regular κ) we can assume that A is a projective algebra such that $|A\restriction a| = \kappa > \aleph_0$ for all non-zero $a \in A$. Let S be an additive open skeleton of A, which exists by 1.3.2(4). The main work will be with the following

CLAIM   For each $S \in \mathcal{S}$ and $a \in A$, if $|S| < |A|$, then $a \in T$ for some $T \in \mathcal{S}$ such that T is countably generated over S and the embedding $S \leq T$ is co-complete with a free one.

Let S be given. First we take $S \leq T_0 \in \mathcal{S}$ such that $a \in T_0$. By additivity, it is possible to choose T_0 countably generated over S. Let M_0 be a maximal disjoint subset of T_0 such that for each $m \in M_0$ there is an element $c_m^0 \in A\restriction m$ which is independent of $T_0 \restriction m$. The existence of M_0 follws from the above lemma. By

---
[3] Here we use an argument suggested by S. Koppelberg, which is easier than our original one.

the ccc, M_0 is countable, so we can find $T_1 \in \mathcal{S}$, which is countably generated over T_0 and such that $\{c_m^0 : m \in M_0\} \subseteq T_1$.

Clearly $|T_1| = |T_0| = |\mathcal{S}| < |A|$. This makes it possible to apply the same procedure to T_1 yielding M_1, $\{c_m^1 : m \in M_1\}$, and then T_2. Continuing in the same way we get a sequence $\mathcal{S} \leq T_0 \leq T_1 \leq T_2 \leq \ldots T_n \leq \ldots$.

Obviously, $T = \bigcup_{n < \omega} T_n$ belongs to $\mathcal{S}$, contains a and is countably generated over $\mathcal{S}$. To see that $\mathcal{S} \leq T$ is co-complete with a free extension, we apply the test 5.2.8.

Only condition $(*)$ requires verification. Let a non-zero $t \in T$ be given. Then $t \in T_n$ for some n. So $t \wedge m > 0$ for some $m \in M_n$. As $c_m^n \in T_{n+1} \leq T$ is independent of $T_n \lceil m$, we get that $t_0 = t \wedge m \wedge c_m^n$ and $t_1 = t \wedge m - c_m^n$ are non-zero.

To verify $\mathcal{S}\lceil - t_0 \subseteq \mathcal{S}\lceil - t_1$, consider any $s \in \mathcal{S}\lceil - t_0$. As $0 \neq s \wedge t_1 = (s \wedge t \wedge m) - c_m^n$ would yield $s \wedge t \wedge m \neq 0$, hence $0 \neq s \wedge t \wedge m \wedge c_m^n = s \wedge t_1$, by independence, it cannot be true. So, $s \wedge t_1 = 0$, i.e. $s \in \mathcal{S}\lceil - t_1$. The other inclusion is established symmetrically.

The claim is proved. From it we get the assertion as in the proof of 1.4.11, using 5.2.7 instead of 1.4.4. □

Together with the proposition, theorem 5.2.2 is now also completely proved.

5.3 Characterizations of weakly projective algebras

The results of this section should be compared with 1.3.2. We find it convenient to list the conditions characterizing weakly projective algebras in two theorems. The first is the basic one.

Theorem 5.3.1 (cf. Theorems 2 and 3 of [49]) *The following assertions are equivalent for any Boolean algebra B.*

(1) *B contains a dense projective subalgebra.*

(2) *B is weakly projective.*

(3) *B has a well-ordered regular filtration $(B_\alpha)_{\alpha < \kappa}$ such that $B_0 = 2$ and each $B_{\alpha+1}$ is countably generated over B_α.*

(4) *B has a well-ordered regular filtration $(B_\alpha)_{\alpha < \kappa}$ such that $B_0 = 2$ and $\pi(B_{\alpha+1}/B_\alpha) = \aleph_0$, for all $\alpha < \kappa$.*

The implication $(1) \Rightarrow (2)$ follows from 5.1.3. $(3) \Rightarrow (4)$ is obvious. The proofs of the other two implications will cost more work. We start with some preparations for $(2) \Rightarrow (3)$.

For a while we fix an embedding $A \leq C$ and say that a subalgebra A' of A is *full* in a subalgebra C' of C (notation $A' \leq_f C'$) if $A' \leq C'$ and for each $c' \in C'$ and all non-zero $a \in A\lceil c'$ there exists some $a' \in A'\lceil c'$ such that $a' \wedge a \neq 0$.

Here are the properties we need.

Lemma 5.3.2 *Assume* $A \leq_d C$.

(1) $A' \leq_f C'$ *implies* $A' \leq_d C'$.

(2) *If A satisfies the ccc, then* $\{S \leq C : S \cap A \leq_f S\}$ *contains a skeleton of C.*

(3) $A' \leq_f C'$ *and* $A'' \leq_f C''$ *imply* $\langle A' \cup A'' \rangle \leq_f \langle C' \cup C'' \rangle$.

(4) *If* $A' \leq_f C'$, *then* $A' \leq_{reg} A$ *iff* $C' \leq_{reg} C$.

Notice that assertions (2) and (4) are also true for dense embeddings. But (3) is not and that is why we introduce full ones.

Proof. (1) should be obvious from the definition of $\leq_f$ and the assumed density of A in C. To prove (2), we fix for each $c \in C$ a maximal subset $F(c)$ of $A \restriction c$ consisting of pairwise disjoint elements. Then each $a \in A \restriction c$ intersects some element of $F(c)$. So if $S \leq C$ is closed under F, then $A \cap S \leq_f S$. By the ccc, $F(c)$ is always countable. It follows that the mapping $F : C \to [C]^{\leq \aleph_0}$ gives rise to a fixed-point skeleton, which is as required.

To prove (3), we consider an arbitrary element $d \in \langle C' \cup C'' \rangle$ and write it as $d = \bigvee_{i=1}^{n} c_i' \wedge c_i''$. Consider $a \in A \restriction d$. There is nothing to do if $a = 0$. Otherwise we can assume $a \wedge c_1' \wedge c_1'' > 0$. Using $A \leq_d C$ we can choose a non-zero $a_1 \in A$ such that $a_1 \leq a \wedge c_1' \wedge c_1''$. Now we use $A' \leq_f C'$ to choose $a' \in A' \restriction c_1'$ such that $a' \wedge a_1 > 0$. As $a' \wedge a_1 \leq a_1 \in A \restriction c_1''$, we find $a'' \in A'' \restriction c_1''$ such that $a'' \wedge a' \wedge a_1 > 0$. Then $a' \wedge a'' \in \langle A' \cup A'' \rangle \restriction (c_1' \wedge c_1'') \subseteq \langle A' \cup A'' \rangle \restriction d$ and $(a' \wedge a'') \wedge a \geq a' \wedge a'' \wedge a_1 > 0$, as desired.

To prove (4) we consider the diagram

$$
\begin{array}{ccc}
A & \leq_d & C \\
\text{VI} & & \text{VI} \\
A' & \leq_d & C',
\end{array}
$$

where the density in the lower row comes from (1). If $C' \leq_{reg} C$, then $A' \leq_{reg} A$, by transitivity. Assume $A' \leq_{reg} A$ and consider $c \in C$. By transitivity, $A' \leq_{reg} C$, which allows us to take some non-zero $a' \in A'$ such that $A' \restriction (a' - c) = \{0\}$. As A' is dense in C' we must have $C' \restriction (a' - c) = \{0\}$, too. □

With the lemma at our disposal we are now ready to **prove** the implication $(2) \Rightarrow (3)$ of the theorem. If B is co-complete to a projective Boolean algebra we can without loss (cf. 5.1.4) assume

$$B \leq_d C \geq_d A,$$

where A is projective. Let S denote an additive open skeleton of A, which exits by 1.3.2. Denote $|B|$ by κ and let $\{b_\alpha : \alpha < \kappa\}$ be an enumeration of B. (We assume that κ is uncountable; otherwise the result is trivial.)

We now construct three continuous chains $(A_\alpha)_{\alpha < \kappa}$, $(B_\alpha)_{\alpha < \kappa}$, and $(C_\alpha)_{\alpha < \kappa}$ of subalgebras of A, B, and C, respectively, such that $A_0 = B_0 = C_0 = 2$ and the following conditions are satisfied for all α.

(1a) $b_\alpha \in B_{\alpha+1}$.

(2a) $B_{\alpha+1}$ is countably generated over B_α.

(3a) $B_\alpha \leq_f C_\alpha$ (with respect to $B \leq_d C$).

(4a) $A_\alpha \leq_f C_\alpha$ (with respect to $A \leq_d C$).

(5a) $A_\alpha \in \mathcal{S}$.

The chain $(B_\alpha)_{\alpha < \kappa}$ will be as required in assertion (3) of the theorem. By (1a) it is a filtration of B. Regularity follows from assertion (4) of the above lemma applied twice to the diagram

$$B \quad \leq_d \quad C \quad \geq_d \quad A$$
$$\mathsf{VI} \qquad \mathsf{VI} \qquad \mathsf{VI}^{reg}$$
$$B_\alpha \quad \leq \quad C_\alpha \quad \geq \quad A_\alpha.$$

For $A_\alpha \in \mathcal{S}$ implies $A_\alpha \leq_{rc} A$ hence $A_\alpha \leq_{reg} A$.

The START of the construction is determined: $A_0 = B_0 = C_0 = 2$. To get continuity, we have to take unions at LIMIT STEPS, which is easily seen to preserve all conditions.

Assuming that A_α, B_α and C_α are given, we now explain the SUCCESSOR STEP of the construction.

CLAIM   *There is a countable subalgebra $X \leq C$ such that $b_\alpha \in X$ and $X \cap A \leq_f X$, $X \cap B \leq_f X$ and $X \cap A \in \mathcal{S}$.*

Indeed, being projective B satisfies the ccc. By 5.1.5, so do A and C. Hence, by (2) of the lemma and 1.1.4, the sets

$$\{T \leq C : T \cap A \leq_f T\}, \quad \{T \leq C : T \cap B \leq_f T\}, \text{ and } \{T \leq C : T \cap A \in \mathcal{S}\}$$

all contain skeletons of C. Their intersection is also a skeleton of C which shows the existence of X.

It remains to put $A_{\alpha+1} = \langle A_\alpha \cup (X \cap A) \rangle$, $B_{\alpha+1} = \langle B_\alpha \cup (X \cap B) \rangle$, and $C_{\alpha+1} = \langle C_\alpha \cup X \rangle$. Conditions (1a) and (2a) are obvious. (3a) and (4a) follow from assertion (3) of the lemma. Finally condition (5a) follows from the additivity of $\mathcal{S}$.

The proof of (2) $\Rightarrow$ (3) is complete. We turn to the implication (4) $\Rightarrow$ (1) starting with a more or less technical

Lemma 5.3.3 *If $A \leq_d B \leq_{reg} C$, then $\langle A \cup X \rangle \leq_d \langle B \cup X \rangle$, for each $X \subseteq C$.*

Proof. There is no loss in assuming that X is in fact a subalgebra of C. So every element of $\langle B \cup X \rangle$ can be written as a finite union of elements of the form $b \wedge x$ for $b \in B$ and $x \in X$. So to prove the lemma it is sufficient to find, for any given $b \wedge x > 0$, some $a \in A$ such that $0 < a \wedge x \leq b \wedge x$. First use $B \leq_{reg} C$ to choose a non-zero $b' \in B$ such that $B \restriction (b' - (b \wedge x)) = \{0\}$. Then $b' \wedge b > 0$

hence $0 < a \leq b' \wedge b$ for some $a \in A$. From $a \wedge x = 0$ we would get $a \leq b' \wedge b - x$, hence $a \in B \lceil (b' - (b \wedge x))$, hence $a = 0$. So $a \wedge x > 0$, as desired. $\square$

Outline of the **proof** of (4) $\Rightarrow$ (1). Starting from the filtration $(B_\alpha)_{\alpha < \kappa}$ of B provided by (4), we construct a continuous chain $(A_\alpha)_{\alpha < \kappa}$ of subalgebras of B such that $A_0 = B_0 = 2$ and the following conditions are satisfied for all $\alpha < \kappa$.

(1b) $A_\alpha \leq_{rc} A_{\alpha+1}$.

(2b) $w(A_{\alpha+1}/A_\alpha) = \aleph_0$.

(3b) $A_\alpha \leq_d B_\alpha$.

If this chain is constucted, then $A = \bigcup_{\alpha < \kappa} A_\alpha$ is the desired subalgebra. By (3b) it is dense and the other conditions yield projectivity, by 1.3.2.

As A_0 is given and continuity leaves no choice in limit steps, we have to care about successor steps only. Assume that $A_\alpha \leq_d B_\alpha$ has been constructed.

CLAIM $\pi(B_{\alpha+1}/A_\alpha) = \aleph_0$

Indeed, condition (2a) allows us to take a countable subalgebra X of $B_{\alpha+1}$ such that $\langle B_\alpha \cup X \rangle$ is dense in $B_{\alpha+1}$. By the lemma, $\langle A_\alpha \cup X \rangle$ is also dense $B_{\alpha+1}$.

Being projective, A_α satisfies the ccc. Therefore the existence of $A_{\alpha+1}$ follows from the claim and the following result that has some interest in its own. It strengthens lemma 1 of [49] (where A had to be projective).

Proposition 5.3.4 *Assume $A \leq_{reg} B$, where $\pi(B/A) = \aleph_0$ and A satisfies the ccc. Then there exits a subalgebra C of B which is countably generated over A and such that $A \leq_{rc} C \leq_d B$.*

Proof. Let $\{b_n : n < \omega\}$ be a subset of B such that $\langle A \cup \{b_n : n < \omega\} \rangle \leq_d B$. We construct an increasing sequence $(C_n)_{n < \omega}$ of subalgebras of B such that $C_0 = A$ and for all $n < \omega$

(1c) $C_n \leq_{rc} C_{n+1} \leq_{reg} B$,

(2c) $C_{n+1} \leq_d C_n(b_n)$,   and

(3c) C_{n+1} is countably generated over C_n.

The algebra $C = \bigcup_{n < \omega} C_n$ then satisfies all requirements: $A = C_0 \leq_{rc} C$, by transitivity (1.2.9), $w(C/A) = \aleph_0$, by construction. To show $C \leq_d B$ it is sufficient to verify $C \leq_d \langle A \cup \{b_n : n < \omega\} \rangle$ for which it is, in turn, sufficient to check $C_n \leq_d A(b_0, b_1, \ldots, b_{n-1})$ [4] for all n. This is clear for $n = 0$. From the inductive hypothesis we get $C_n(b_n) \leq_d A(b_0, b_1, \ldots, b_{n-1}, b_n)$, by lemma 5.3.3. Hence $C_{n+1} \leq_d A(b_0, b_1, \ldots, b_{n-1}, b_n)$, by (2c).

It remains to actually construct the sequence $(C_n)_{n < \omega}$. As C_0 is given, we have to explain how C_{n+1} can be constructed from C_n. To get rid of indices, we formulate an extra

---

[4] Recall that $A(b_1, \ldots, b_n)$ is shorthand for $\langle A \cup \{b_1, \ldots, b_n\} \rangle$.

Lemma 5.3.5 *Assume* $A \leq_{reg} B$, *where* A *satisfies the ccc. Then for each* $b \in B$ *there exists a subalgebra* C *of* B *which is countably generated over* A *and such that* $A \leq_{rc} C \leq_d A(b) \leq_{reg} B$.

In the proof of this lemma we need the folowing trivial remark. Not to get puzzeled then, we separate it beforehand. It can be verified using disjunctive normal form, for example.

Observation 5.3.6 *If the Boolean algebra* C *is generated by its subset* G *which is closed under intersections and such that all complements of elements of* G *are finite unions of elements of* G, *then each element of* C *is a finite union of elements of* G.

Proof of the lemma. Let $b \in B$ be given. The assertion $A(b) \leq_{reg} B$ has nothing to do with C and is true by 1.2.19. To construct the required C we consider the ideal

$$I = (A{\upharpoonright} b)^d \cap (A{\upharpoonright} - b)^d.$$

In the MAIN CASE $I \neq \{0\}$, we choose a maximal subset of I consisting of pairwise disjoint non-zero elements. By the ccc, this set can be enumerated (with repetitions if it is finite) as $\{a_n : n < \omega\}$.

We let C be the subalgebra of B generated by

$$A \cup \{a_n \wedge b : n < \omega\} \cup \{a_n - b : n < \omega\}.$$

Obviously, C is countably generated over A and contained in $A(b)$. Let us show that C is actually dense in $A(b)$.

As each element of $A(b)$ can be written in the form $(a' \wedge b) \vee (a'' - b)$ for suitable $a', a'' \in A$ and as b and $-b$ occur symmetrically in the construction of C, our task reduces to finding a non-zero $c \in C$ below any given $a \wedge b > 0$ with $a \in A$. The regularity of A in B yields a non-zero element $a' \in A$ such that $A{\upharpoonright} (a' - (a \wedge b)) = \{0\}$. Clearly, $0 < a' \wedge a \wedge b \leq a' \wedge a$. It also follows that $A{\upharpoonright} (a' \wedge a - b) \subseteq A{\upharpoonright} (a' - (a \wedge b)) = \{0\}$, which can be reformulated as

(1) $0 \neq a' \wedge a \in (A{\upharpoonright} - b)^d$.

Now we distinguish two subcases. It may be that there is some $a'' \in A$ such that $a'' \leq b$ and $a'' \wedge a' \wedge a \neq 0$. In this case $c = a'' \wedge a' \wedge a$ is as required and even better because it belongs to A and not only to C.

To say that there is no such a'' means to say $a' \wedge a \in (A{\upharpoonright} b)^d$. From (1) it follows that in this subcase, $a' \wedge a$ belongs to I. The maximality of the sequence chosen at the beginning yields $a' \wedge a \wedge a_n \neq 0$, for some $n < \omega$. We claim that $c = a' \wedge a \wedge (a_n \wedge b)$ has the desired properties. Indeed, $c \in C$ and $c \leq a \wedge b$ are obvious. If c were 0, then $a' \wedge a \wedge a_n \leq -b$, i.e. $a' \wedge a \wedge a_n \in A{\upharpoonright} - b$. As $a_n \in (A{\upharpoonright} - b)^d$ we would get the contradiction $0 = a_n \wedge (a' \wedge a \wedge a_n) = a' \wedge a \wedge a_n \neq 0$.

In both cases we have found the desired element $c \in C$; so $C \leq_d A(b)$ is proved.

We still have to establish $A \leq_{rc} C$. Before starting the work proper, we describe the elements of C. For each $a \in A \upharpoonright a_n$, the elements $a \wedge b = a \wedge (a_n \wedge b)$ and $a - b = a \wedge (a_n - b)$ belong to C. It follows that the set

$$G = A \cup \{a \wedge b : a \in A \upharpoonright a_n \text{ for some } n\} \cup \{a - b : a \in A \upharpoonright a_n \text{ for some } n\}$$

is contained in C. As G includes all the generators mentioned in the definition of C, it generates C. It is easy to see that G is closed under intersections. Moreover, each complement of an element of G is a finite union of elements of G. For example, if $a \leq a_n$, then $-(a \wedge b) = -a_n \vee (a_n - b) \vee ((a_n - a) \wedge b)$. So by the remark preceeding the lemma, each element of C is a finite union of elements of G. It now follows from 1.2.3 that in order to check $A \leq_{rc} C$ it is sufficient to fing a minimal element of A above each $g \in G$. This is trivial for $g \in A$ and equally difficult for the two cases $g = (a \wedge b)$ and $g = (a - b)$. We confine ourselves to the first one. Consider some $a \leq a_n$. We claim that a is the required minimal element of A above $a \wedge b$. Indeed, $a \wedge b \leq a$ is clear. So assume $a \wedge b \leq a' \leq a_n$ for some $a' \in A$. Then $(a - a') \wedge b = 0$, i.e. $a - a' \in A \upharpoonright -b$. As $a_n \in (A \upharpoonright -b)^d$ we get $0 = a_n \wedge (a - a') = a - a'$, i.e. $a \leq a'$, as desired. The main case is finished.

In the SPECIAL CASE $I = \{0\}$ the algebra A itself can be taken as C. Indeed, $A \leq_{rc} A$ is trivial and $A \leq_d A(b)$ can be proved as $C \leq_d A(b)$ in the main case, only that the second subcase can no longer occur.

The proof of the lemma is complete. It yields the proposition and, thereby, the missing implication of the theorem. □

Before giving the second bunch of characterizations of weakly projective algebras we pause a little to care for embeddings that will turn out to form the adequate class. It is natural to expect that these will be the embeddings that are co-complete with projective ones. We call them *weakly projective* and write $A \leq_{wproj} B$. The following characterizations are very much in parallel to the previous theorem and also to 1.5.2.

Theorem 5.3.7 *Assume that $B' \leq B$ and that B' satisfies the ccc. Then the following assertions are equivalent.*

(1) *There exists a dense subalgebra C of $\overline{B}$ such that $B' \leq_{proj} C$.*

(2) $B' \leq_{wproj} B$.

(3) *B has a well-ordered regular filtration $(B_\alpha)_{\alpha < \kappa}$ such that $B_0 = B'$ and each $B_{\alpha+1}$ is countably generated over B_α.*

(4) *B has a well-ordered regular filtration $(B_\alpha)_{\alpha < \kappa}$ such that $B_0 = B'$ and $\pi(B_{\alpha+1}/B_\alpha) = \aleph_0$, for all $\alpha < \kappa$.*

Proof. The implications $(1) \Rightarrow (2)$ and $(3) \Rightarrow (4)$ are obvious.

$(2) \Rightarrow (3)$. By definition, $B' \leq_{wproj} B$ allows us to assume the following situation

$$
\begin{array}{ccccc}
B & \leq_d & C & {}_d\geq & A \\
\vert\vert & & \vert\vert & & \vert\vert^{proj} \\
B' & \leq_d & C' & {}_d\geq & A'.
\end{array}
$$

The required filtration will be constructed exactly as in the proof of the implication $(2) \Rightarrow (3)$ of 5.3.1, starting from an additive rc-skeleton of A over A', which exists by 1.5.2(3). The only difference is that the construction starts with $A_0 = A'$, $B_0 = B'$, and $C_0 = C'$. There are two conditions that require additional checking, namely

(i) C and all its subalgebras satisfy the ccc, and

(ii) A' and B' are both full in C'.

To verify (i), notice that B' satisfies the ccc, by assumption. It remains to notice that, by 1.5.6, the ccc is preserved going through

$$ B' \leq_d C' \geq A' \leq_{proj} A \leq_d C. $$

To verify $B' \leq_f C'$, let $c' \in C'$ and a non-zero $b \in B \vert c'$ be given. Using regularity (by 5.1.9), we can choose a non-zero $b' \in B'$ such that $B' \vert (b' - b) = \{0\}$. Then $b' \leq c'$, for, otherwise, density of B' in C' yields a non-zero $b'' \leq b' - c'$, which would belong to $B' \vert (b' - c') \subseteq B' \vert (b' - b) = \{0\}$. Moreover, $b' \wedge b \neq 0$, as required.

$A' \leq_f C'$ is verified in the same way. It is a little easier because we can use $A' \leq_{rc} A$, by 1.5.6.

The final implication $(4) \Rightarrow (1)$ is proved as in 5.3.1, except that the parallel chain $(A_\alpha)_{\alpha < \kappa}$ starts with B'. The only point that needs additional verification is the ccc for all A_α. But it is assumed for $A_0 = B'$ and each A_α is, by construction, a projective extension of A_0, so the ccc is preserved, by 1.5.6. $\square$

Now we take up the subject of characterizing weakly projective algebras again. In the proof of the second characterization theorem the following little result is needed.

Lemma 5.3.8 *If $\varphi : A \to A$ is a homomorphism such that $\{c \in A : \varphi(c) = c\}$ is dense in A, then $\varphi = id$.*

Proof. Assuming the contrary, we get some a such that $\varphi(a) \neq a$. We can assume $a - \varphi(a) > 0$, for, otherwise, $-a - \varphi(-a) = \varphi(a) - a > 0$ and we replace a by $-a$. By assumption, there exists some $c > 0$ such that $c \leq a - \varphi(a)$ and $\varphi(c) = c$. But then, $c \leq a$ implies $c = \varphi(c) \leq \varphi(a)$. On the other hand, $c \leq -\varphi(a)$, so $0 < c \leq \varphi(a) \wedge -\varphi(a)$, which is the desired contradiction. $\square$

Theorem 5.3.9 *The following assertions are equivalent for any Boolean algebra*
B.

(1) *B is weakly projective.*

(2) *For each homomorphism $\varphi : \overline{A} \to \overline{B}$ such that $\varphi(A) = B$ there exists a*
homomorphism $\varepsilon : \overline{B} \to \overline{A}$ such that $\varphi \circ \varepsilon = id_{\overline{B}}$ and for all non-zero $b \in B$
there is some $a \in A$ such that $a \leq \varepsilon(b)$ and $0 < \varphi(a)$.

(3) *For each surjective homomorphism $\varphi : A \to B$ there is a mapping δ which*
to each $b \in B$ attaches a regular ideal of A such that

  (3i) $\delta(0) = \{0\}$,

  (3ii) $\delta(b_1 \wedge b_2) = \delta(b_1) \cap \delta(b_2)$, *and*

  (3iii) $\varphi(\delta(b))$ *is a dense subset of $B \upharpoonright b$.*

(4) *There are a free Boolean algebra F, a surjective homomorphism $\varphi : F \to B$*
and a mapping δ which to each $b \in B$ attaches an ideal of F such that
conditions (3i), (3ii), and (3iii) are satisfied.

(5) *B has an additive regular skeleton.*

(6) *(Jech, unpublished) There is a club C of countable regular subalgebras of*
B such that $\langle C_1 \cup C_2 \rangle \in C$, for all $C_1, C_2 \in C$.

(7) *(Koppelberg, unpublished; Bandlow, implicit in [4]) There exists a fam-*
ily $(B_j)_{j \in J}$ of countable subalgebras of B such that $B = \bigcup_{j \in J} B_j$ and
$\langle \bigcup_{j \in K} B_j \rangle \leq_{reg} B$, for all $K \subseteq J$.

Proof. (1) $\Rightarrow$ (2). Let $\varphi : \overline{A} \to \overline{B}$ be given such that $\varphi(A) = B$. By 5.3.1(1),
there exists a dense projective subalgebra C, say, of B. Denote $\varphi^{-1}(C) \cap A$ by
D and let $\psi : D \to C$ denote the restriction of φ to D considered as a mapping
into C. The situation is illustrated by the follwing diagram.

$$
\begin{array}{ccccc}
D & \leq & A & \leq_d & \overline{A} \\
\downarrow \psi & & \downarrow \varphi & & \downarrow \varphi \\
C & \leq_d & B & \leq_d & \overline{B}.
\end{array}
$$

From $\varphi(A) = B$ it follows that ψ is surjective. So, projectivity yields a homo-
morphism $\alpha : C \to D$ such that $\psi \circ \alpha = id_C$. Let $\varepsilon : \overline{B} \to \overline{A}$ be a lifting of α.
As $\varphi \circ \varepsilon$ coincides with $\psi \circ \alpha = id_C$ on a dense subalgebra of $\overline{B}$, we must have
$\varphi \circ \varepsilon = id_{\overline{B}}$, by the above little lemma.

To check the extra condition, consider any non-zero $b \in B$. Take $c \in C$ such
that $0 < c \leq b$. Then $a = \varepsilon(c) = \alpha(c) \in D \subseteq A$ satisfies $a = \varepsilon(c) \leq \varepsilon(b)$ and
$\varphi(a) = c > 0$, i.e. a is as desired.

(2) $\Rightarrow$ (3). Let $\varphi : A \to B$ be any surjective homomorphism and $\overline{\varphi} : \overline{A} \to \overline{B}$
some lifting. As $\overline{\varphi}(A) = \varphi(A) = B$, condition (2) yields $\varepsilon : \overline{B} \to \overline{A}$. We put

$$
\delta(b) = \{c \in A : c \leq \varepsilon(b)\} = A \upharpoonright \varepsilon(b)
$$

and show that this mapping works. As A is dense, hence regular, in $\overline{A}$, each $\delta(b)$ is, by construction, a regular ideal of A.

It is clear that $\delta(0) = \{0\}$. Moreover,

$$
\begin{aligned}
\delta(b_1 \wedge b_2) &= \{c \in A : c \leq \varepsilon(b_1 \wedge b_2) = \varepsilon(b_1) \wedge \varepsilon(b_2)\} \\
&= \{c \in A : c \leq \varepsilon(b_1)\} \cap \{c \in F : c \leq \varepsilon(b_2)\} \\
&= \delta(b_1) \cap \delta(b_2).
\end{aligned}
$$

From $A \ni c \leq \varepsilon(b)$ it follows that $\varphi(c) = \overline{\varphi}(c) \leq \varphi(\varepsilon(b)) = b$. So $\varphi(\delta(b)) \subseteq B\!\restriction b$. It remains to see that $\varphi(\delta(b))$ is actually dense in $B\!\restriction b$. So let some non-zero $b' \leq b$ be given. The 'extra condition' in (2) yields $c \in A$ such that $c \leq \varepsilon(b')$ and $\varphi(c) > 0$. From $\varepsilon(b') \leq \varepsilon(b)$ we get $c \in \delta(b)$ and from $c \leq \varepsilon(b')$ we have $\varphi(c) = \overline{\varphi}(c) \leq \varphi(\varepsilon(b')) = b'$, as desired.

(3) $\Rightarrow$ (4) is trivial. To prove (4) $\Rightarrow$ (5), let $\varphi : F \to B$ and δ be as in (4). We start with the following

CLAIM 0. *The mapping* $b \mapsto \delta(b)^{dd}$ *also satisfies conditions* $(3i) - (3iii)$.

Indeed, $\delta(0)^{dd} = \{0\}^{dd} = \{0\}$ is obvious. From lemma 1.2.13 we get

$$
\delta(b_1 \wedge b_2)^{dd} = (\delta(b_1) \cap \delta(b_2))^{dd} = \delta(b_1)^{dd} \cap \delta(b_2)^{dd}.
$$

As $\delta(b) \subseteq \delta(b)^{dd}$, we get $(3iii)$ if we verify $\varphi(\delta(b)^{dd}) \subseteq B\!\restriction b$. If this were not the case, then $b' = \varphi(c) - b > 0$ for some $c \in \delta(b)^{dd}$.

As $\varphi(\delta(b'))$ is dense in $B\!\restriction b'$, we can take some $x \in \delta(b') \subseteq \delta(b')^{dd}$ such that $0 < \varphi(x) \leq b' \leq \varphi(c)$, which implies $x \wedge c \neq 0$.

On the other hand, from $\delta(b') \cap \delta(b) = \delta(b' \wedge b) = \delta(0) = \{0\}$, we get $x \wedge c \in \delta(b')^{dd} \cap \delta(b)^{dd} = \{0\}$, which is the desired contradiction. Claim 0 is proved. It allows us to assume in the rest of the proof that all ideals $\delta(b)$ are regular.

From now on the argument parallels and should be compared with the proof of the implication (3) $\Rightarrow$ (4) of theorem 1.3.2.

Let U be any set of free generators of F. We call a subset V of U *admissible* if for every $w \in V \cup -V$ the regular ideal $\delta(\varphi(w))$ is generated by its intersection with $\langle V \rangle$

CLAIM 1. *Every subset of* U *is contained in an admissible subset of the* *same cardinality.*

Indeed, for each $u \in U$ we can fix a countable set $f(u) \subseteq U$ such that $\delta(\varphi(u))$ and $\delta(\varphi(-u))$ are both generated by their intersection with $\langle f(u) \rangle$. This is because the ideals are regular and F has the (BSP). Now any given $W \subseteq U$ may be 'closed under f', which yields the claim.

CLAIM 2. *The union of every collection of admissible sets is admissible.*

This is obvious. As in the old proof, the two claims yield that

$$S = \{\varphi(\langle V \rangle) : V \text{ is an admissible subset of } U\}$$

is an additive skeleton of B. It remains to verify that S consists of regular subalgebras of B, i.e. to establish

CLAIM 3. *If $V \subseteq U$ is admissible, then $\varphi(\langle V \rangle) \leq_{reg} B$.*

Let a non-zero $b \in B$ be given. We must find a non-zero $a \in \varphi(\langle V \rangle)$ such that

$$\varphi(\langle V \rangle) \restriction (a - b) = \{0\}.$$

By $(3iii)$, $\varphi(\delta(b))$ is dense in $B \restriction b$, which allows us to take $c \in \delta(b) \subseteq F$ such that $0 < \varphi(c) \leq b$. Let d be minimal in $\langle V \rangle \leq_{rc} F$ above c. We show that $a = \varphi(d)$ is as desired.

Clearly, $a = \varphi(d) \geq \varphi(c) > 0$. Aiming at a contradiction we now assume $\varphi(\langle V \rangle) \restriction (a - b) \neq \{0\}$ and pick an element $x \in \langle V \rangle$ such that

$$(*) \quad 0 < \varphi(x) \leq a = \varphi(d) \quad \text{and} \quad (**) \quad \varphi(x) \wedge b = 0.$$

We can assume that $x = w_1 \wedge \ldots \wedge w_n$, where the w_i belong to $V \cup -V$. Indeed, the first choice for x is a join of elements of this form and one of the joinands can replace x.

SUBCLAIM *There exists $y \in \langle V \rangle$ such that $y \in \delta(\varphi(x))$ and $\varphi(y) > 0$.*

As $\varphi(\delta(\varphi(x)))$ is dense in $B \restriction \varphi(x) \neq \{0\}$, there must be some $z \in \delta(\varphi(x))$ such that $\varphi(z) > 0$. Condition $(3ii)$ and the special form of x, yield $z \in \delta(\varphi(x)) = \delta(\varphi(w_1)) \cap \ldots \cap \delta(\varphi(w_n))$. From $z \in \delta(\varphi(w_i))$ and admissibiltiy of V we get some $y_i \in \langle V \rangle$ such that $z \leq y_i \in \delta(\varphi(w_i))$. Putting $y = \bigwedge_{i=1}^{n} y_i$ yields the sought for y. Indeed, $y \leq y_i$ implies $y \in \delta(\varphi(w_i))$ for all i, hence

$$y \in \delta(\varphi(w_1)) \cap \ldots \cap \delta(\varphi(w_n)) = \delta(\varphi(x)).$$

From $z \leq y$ we also get $\varphi(y) \geq \varphi(z) > 0$. The subclaim is proved.

As $y \in \delta(\varphi(x))$, we must have $\varphi(y) \leq \varphi(x)$, which yields $\varphi(y) \leq \varphi(d)$, by $(*)$. So $\varphi(y \wedge d) = \varphi(y) > 0$, hence $y \wedge d \neq 0$. It follows that $y \wedge c \neq 0$, because d was minimal above c in $\langle V \rangle$. Consequently,

$$0 \neq y \wedge c \in \delta(\varphi(x)) \cap \delta(b) = \delta(\varphi(x) \wedge b).$$

By $(3i)$, $\varphi(x) \wedge b \neq 0$, contradicting $(**)$. Claim 3 is proved and thereby the implication $(4) \Rightarrow (5)$ of the theorem.

$(5) \Rightarrow (6)$. The club consists of all countable members of the additive regular skeleton.

$(6) \Rightarrow (7)$. Let $\mathcal{C}$ be the club in question. First notice that the existence of $\mathcal{C}$ implies the ccc. This follows from a by now familiar argument using 1.1.6, 5.3.1(3), and 5.1.5.

It will be sufficient to show that, actually, $\langle \bigcup \mathcal{K} \rangle \leq_{reg} B$ for all $\mathcal{K} \subseteq \mathcal{C}$. For finite $\mathcal{K}$ this follows from the assumption. If $\mathcal{K} = \{K_1, K_2, \ldots\}$ is countable, then $\langle \bigcup \mathcal{K} \rangle = \bigcup_{n=1}^{\infty} \langle K_1 \cup \ldots \cup K_n \rangle$ belongs to $\mathcal{C}$ and is, therefore, regular in B.

Now let $\mathcal{K}$ be uncountable. We use 1.2.16(5) to test $\langle \bigcup \mathcal{K} \rangle \leq_{reg} B$. Let M be a maximal disjoint subset of $\langle \bigcup \mathcal{K} \rangle$. We have to show that M is maximal in B. By the ccc, M is countable. So $M \subseteq \langle \bigcup \mathcal{L} \rangle$, for some countable $\mathcal{L} \subseteq \mathcal{K}$. Clearly, M is maximal in $\langle \bigcup \mathcal{L} \rangle \leq_{reg} B$, hence in B.

The implication $(7) \Rightarrow (1)$ is easily established, using condition 5.3.1(3). The theorem is proved. $\square$

In connection with the implication $(1) \Rightarrow (2)$ we mention that an even simpler prove (not caring about the extra condition) yields

Observation 5.3.10 *If A is weakly projective, then $\overline{A}$ is a projective object in the full subcategory of complete Boolean algebras.*

It is an **open problem** whether the converse is also true. Shapiro proved in [46] that the answer is affirmative for weakly projective algebras of π-weight $\leq \aleph_1$.

Our first application of the characterization theorems is the, historically, first major result on weakly projective algebras.

Theorem 5.3.11 (Shapiro [49]) *Every subalgebra of a projective Boolean algebra contains a dense projective subalgebra and is, therefore, weakly projective.*

Proof. The formulation is chosen to impress the superficial reader. In the proof below we use 5.3.9(3) to establish that the algebra is weakly projective. The dense projective subalgebra is then automatically provided by 5.3.1(1). Moreover, the proof immediately boils down to subalgebras of free Boolean algebras, given that every projective Boolean algebra is a retract, hence subalgebra, of a free one.

So let $A \leq F$ be given, where F is free. We begin by choosing an ideal I of F which is maximal with respect to the property that $A \cap I = \{0\}$. Let us denote F/I by B and the canonical mapping $F \to F/I$ by φ. By 1.2.21, $\varphi \restriction A : A \to B$ is an injective homomorphism onto a dense subalgebra of B.

Consequently, $A \sim B$ and our task is reduced to proving B co-complete to a projective Boolean algebra. In order to apply 5.3.9(3), we let $\delta(b)$ be the ideal of F generated by $\{a \in A : \varphi(a) \leq b\}$, i.e.

$$\delta(b) = \{f \in F : \text{ there exists } a \in A \text{ such that } f \leq a \text{ and } \varphi(a) \leq b\}.$$

It remains to verify conditions $(3i) - (3iii)$. By injectivity of φ, $\varphi(a) \leq 0$ is true for $a = 0$ only, hence $\delta(0) = \{0\}$.

To check $\delta(b_1 \wedge b_2) = \delta(b_1) \cap \delta(b_2)$ assume first $f \in \delta(b_1 \wedge b_2)$. Then $f \leq a$ and $\varphi(a) \leq b_1 \wedge b_2$ for some $a \in A$. Clearly, the same a witnesses $f \in \delta(b_1)$ and $f \in \delta(b_2)$. The other way round, let $f \in \delta(b_1) \cap \delta(b_2)$ be witnessed by a_1 and a_2, respectively. Then $f \leq a_1$ and $f \leq a_2$ imply $f \leq a_1 \wedge a_2$, and $\varphi(a_1) \leq b_1$

and $\varphi(a_2) \leq b_2$ imply $\varphi(a_1 \wedge a_2) \leq b_1 \wedge b_2$. So $a_1 \wedge a_2$ witnesses $f \in \delta(b_1 \wedge b_2)$, as desired.

If $f \in \delta(b)$, then $f \leq a$ and $\varphi(a) \leq b$, hence $\varphi(f) \leq b$. This shows $\varphi(\delta(b)) \subseteq B \upharpoonright b$.

To get the density, let $0 < b' \leq b$ be given. As $\varphi(A)$ is dense in B, we find $a \in A$ such that $0 < \varphi(a) \leq b'$. As a itself witnesses $a \in \delta(b)$, we are done. $\square$

In connection with the previous theorem we want to draw the readers attention to two **open questions**. The first of them has, in similar form, been asked by B A. Efimov.

> *Is it true that each subalgebra of a projective Boolean algebra can be densely embedded into a projective Boolean algebra?*

The second question concerns weakly projective algebras.

> *If A is weakly projective and all ultrafilters of A have π-character τ, does it follow that A has a dense subalgebra isomorphic to $\mathrm{Fr}\,\tau$?*

It is known (Shapiro, unpublished) that the answer is positive if A is a subalgebra of a free Boolean algebras and, more generally, for algebras of the form $F(A)$, where A is a subalgebra of a free algebra and F a normal functor.

DIGRESSION

Let us mention that some of the notions defined above could also be introduced in the categorical spirit of the beginning of section 1.4. Recall that $\mathcal{B}_A$ is the category of Boolean algebras that contain A with morphisms that keep A point-wise fixed. It makes sense to consider the usual $\overline{B}$ as the completion of an object $B \in \mathcal{B}_A$. Then two objects could be defined co-complete if in $\mathcal{B}_A$ they have isomorphic completions, i.e. the embeddings $A \leq B_1$ and $A \leq B_2$ are co-complete in this sense iff there is an isomorphism $\overline{B_1} \to \overline{B_2}$, which keeps A fixed.

With this in mind, it is natural to call $A \leq B$ co-complete with a free, resp. projective, embedding iff B is co-complete with a free, resp. projective, object in $\mathcal{B}_A$.

It turns out, however, that using this approach we arrive at the same notions as above. Indeed, 5.2.6 and 5.3.7(1) express just this fact.

5.4   Adequateness and closedness properties

First of all we establish the adequatness result promised above.

Theorem 5.4.1 *The class of weakly projective algebras is adequate to the class of weakly projective embeddings.*

The proof is a minor modification of that of 1.5.4, using the above characterization theorems. $\square$

The following observation will be needed below, but it is also interesting in its own right.

Proposition 5.4.2 *A free product $A \otimes B$ is weakly projective iff both factors are weakly projective.*

Proof. If A and B are weakly projective, we can fix dense projective subalgebras $A' \leq A$ and $B' \leq B$. Clearly $A' \otimes B'$ is projective and dense in $A \otimes B$, so the latter is weakly projective.

Leaving the details to the reader, we outline the proof of the other direction. Assuming that $A \otimes B$ is weakly projective one fixes an additive regular skeleton, S say. Intersecting S with $\{A' \otimes B' : A' \leq A; \ B' \leq B\}$, yields a regular skeleton, which is still additive because of the formula

$$\langle \bigcup_{i \in I} (A_i \otimes B_i) \rangle = \langle \bigcup_{i \in I} A_i \rangle \otimes \langle \bigcup_{i \in I} B_i \rangle.$$

The next step is to verify that $A' \otimes B' \leq_{reg} A \otimes B$ implies $A' \leq_{reg} A$ and $B' \leq_{reg} B$. Finally one shows that

$$\{A' \leq A : |A'| = \aleph_0; \ A' \otimes B' \in S \text{ for some } B' \leq B\}$$

is as required in 5.3.9(7). Similarly for B. □

It would be tempting to generalize this result on factors of free products to the assertion that retracts of weakly projective algebras are weakly projective again. But this generalization turns out to be false.

Information 5.4.3 ([46]) *There is a compact and zero-dimensional, separable space, which is not co-absolute to any dyadic space.*

Denote the clopen algebra of this space by A. It is not weakly projective. On the other hand, separability yields an embedding of A into the power-set algebra $P(\omega)$, which, by Sikorski's Theorem, lifts to an embedding of $\overline{A}$ into $P(\omega)$. Again by Sikorski's Theorem, the identical mapping $\overline{A} \twoheadrightarrow \overline{A}$ then lifts to a retraction $P(\omega) \to \overline{A}$. So, $\overline{A}$ is not weakly projective but a retract of the weakly projective algebra $P(\omega)$. □

Now we come to subalgebras. The situation is the same as with projectivity.

Proposition 5.4.4 *Assume that $A \leq_{wproj} B$.*

(1) *If A is weakly projective, then B is weakly projective.*

(2) *If B is weakly projective, then A is weakly projective.*

Proof. Assertion (1) is trivial, take a filtration of A as in 5.3.1(3). Then use 5.3.7(3) to prolong it to a filtration of B.

To prove (2), we choose dense subalgebras C, D of $\overline{B}$ such that D is projective and

$$A \leq_{proj} C \leq_d \overline{B} \geq_d D.$$

This is possible, by 5.3.7(1) and 5.3.1(1). By 1.5.7, there is a free algebra F such that $A \otimes F \simeq C \otimes F$. As $D \otimes F$ is projective,

$$A \otimes F \simeq C \otimes F \leq_d \overline{B} \otimes F \geq_d D \otimes F$$

implies that $A \otimes F$ is weakly projective. Consequently, A is weakly projective. □

In 3.4 we have constructed a relatively complete subalgebra of a projective Boolean algebra, namely $SP^2(\mathrm{Fr}\,\omega_2)$, which is not projective. The parallel result for weakly projective algebras has recently been obtained by Koppelberg and Shelah [37] who constructed a relatively complete subalgebra of a weakly projective algebra, which is not itself weakly projective.

In section 3.3 we have seen that the functor exp does not preserve projectivity. As our second application of the characterization theorems we show that in the class of weakly projective algebras the situation is better.

Theorem 5.4.5 *A Boolean algebra is weakly projective iff its exponential is also weakly projective.*

Proof. One direction is easy because of the preparations we made in section 3.3. Let $(A_\alpha)_{\alpha<\gamma}$ be a filtration as in 5.3.1(3), i.e. $A_0 = 2$, $A_\alpha \leq_{reg} A$ and $w(A_{\alpha+1}/A_\alpha) = \aleph_0$.

By continuity of exp, the chain $(exp\,A_\alpha)_{\alpha<\gamma}$ is a filtration of $exp\,A$. By 3.3.14, $exp\,A_\alpha \leq_{reg} exp\,A$ and, by 3.3.16,

$$\pi(exp\,A_{\alpha+1}/exp\,A_\alpha) \leq \pi(A_{\alpha+1}/A_\alpha) = \aleph_0.$$

Here we need that $\aleph_1$ is a caliber of A and, therefore, of all its subalgebras. This follows from 5.1.5 and 2.7.2. We have checked that the filtration $(exp\,A_\alpha)_{\alpha<\gamma}$ is as demanded in 5.3.1(4), so $exp\,A$ is weakly projective.

The other direction will take some more effort. Assume that $exp\,A$ is weakly projective. By 5.3.9(4), we can fix an additive regular skeleton $\mathcal{S}$, say, of $exp\,A$. We want to show that A is also weakly projective. This being trivial for countable A, we assume that A is uncountable and denote the first ordinal of cardinality $|A|$ by ρ.

Our aim is to construct, by induction on $\alpha < \rho$, a well-ordered regular filtration $(A_\alpha)_{\alpha<\rho}$ of A starting with $A_0 = 2$ such that $w(A_{\alpha+1}/A_\alpha) = \aleph_0$ for all α. By 5.3.1, the existence of this filtration then proves that A is weakly projective.

To keep the induction going, we construct parallel to the A_α elements $S_\alpha \in \mathcal{S}$ such that $(S_\alpha)_{\alpha<\rho}$ becomes a filtration of $exp\,A$, each $S_{\alpha+1}$ is countably generated over S_α, and the crucial condition $S_\alpha \leq_d exp\,A_\alpha$ is satisfied for all α.

We now fix enumerations $\{a_\alpha : \alpha < \rho\}$ of A and $\{h_\alpha : \alpha < \rho\}$ of $exp\,A$, respectively. To guarantee that both sequences are exhausting, we make sure that $a_\alpha \in A_{\alpha+1}$ and $h_\alpha \in S_{\alpha+1}$.

There is no loss in assuming $exp\, 2 = \{0,1\} \in \mathcal{S}$ and we may START with $A_0 = 2$ and $S_0 = exp\, A_0$.

At LIMIT STAGES we have to take unions and this will not destroy any of the conditions, because of the continuity of exp. The real interesting thing happens in the

SUCCESSOR STEP. Assume that we have proceeded until $A_\alpha \leq_{reg} A$ and $S_\alpha \leq_d exp\, A_\alpha$. Put $B_0 = A_\alpha(a_\alpha)$ and $T_0 = S_\alpha$. Notice that the filtration $(A_\beta)_{\beta \leq \alpha}$ constructed so far witnesses that A_α is weakly projective. It follows that $\aleph_1$ is a caliber of A_α. Moreover, $A_\alpha \leq_{reg} A$, by induction hypothesis, so $A_\alpha \leq_{reg} B_0$. From 3.3.16 we therefore get $\pi(exp\, B_0 / exp\, A_\alpha) \leq \pi(B_0/A_\alpha) = \aleph_0$, which allows us to fix a countable subalgebra D_0 of $exp\, B_0$ such that $\langle D_0 \cup exp\, A_\alpha \rangle \leq_d exp\, B_0$.

As $\{h_\alpha\} \cup D_0$ is a countable subset of $exp\, A$, we find $T_1 \in \mathcal{S}$ countably generated over T_0 such that $\{h_\alpha\} \cup D_0 \subseteq T_1$, by the additivity of $\mathcal{S}$.

Next we take $B_1 \leq A$, countably generated over B_0 such that $T_1 \leq exp\, B_1$. To see that this is possible, write $T_1 = \langle T_0 \cup X \rangle$ for some countable X. Then the formal sums $a_1 \oplus a_2 \oplus \ldots \oplus a_n$ making up the elements of X mention only countably many elements of A. Collecting them into the set $Y \subseteq A$, we can put $B_1 = \langle B_0 \cup Y \rangle$.

Now a new round starts. As B_1 is still countably generated over A_α, we have $\pi(exp\, B_1 / exp\, A_\alpha) = \aleph_0$ and can take a countable $D_1 \subseteq exp\, B_1$ such that $\langle D_1 \cup exp\, A_\alpha \rangle \leq_d exp\, B_1$. The set D_1 can be caught into a $T_2 \in \mathcal{S}$ wich is countably generated over T_1, after which we find $B_2 \leq A$ countably generated over B_1 such that $T_2 \leq exp\, B_2$.

Repeating this process over and over again, we arrive at the following chains

$$
\begin{array}{ccc}
S_\alpha & \leq_d & exp\, A_\alpha \\
\| & & \wedge| \\
T_0 & & exp\, B_0 \supseteq D_0 \\
\wedge| & & \wedge| \\
\{h_\alpha\} \cup D_0 \subseteq T_1 & \leq & exp\, B_1 \supseteq D_1 \\
\wedge| & & \wedge| \\
D_1 \subseteq T_2 & \leq & exp\, B_2 \supseteq D_2 \\
\vdots & & \vdots \\
D_n \subseteq T_{n+1} & \leq & exp\, B_{n+1} \supseteq D_{n+1} \\
\vdots & & \vdots
\end{array}
$$

where

$$T_n \in \mathcal{S},\ w(T_{n+1}/T_n) = w(B_{n+1}/B_n) = \aleph_0,\ \text{and}\ \langle D_n \cup exp\, A_\alpha \rangle \leq_d exp\, B_n.$$

Of course, we now put $S_{\alpha+1} = \bigcup_{n < \omega} T_n$ and $A_{\alpha+1} = \bigcup_{n < \omega} B_n$. The other conditions being obvious, we verify the crucial ones:

(1) $S_{\alpha+1} \leq_d exp\, A_{\alpha+1}$ and   (2) $A_{\alpha+1} \leq_{reg} A.$

(Notice here, that condition (2) is our only interest; all the fuss about the S_α is made to secure it.)

To check (1), let a non-zero element $x \in exp A_{\alpha+1} = \bigcup_{n<\omega} exp\, B_n$ be given. Then $x \in exp\, B_n$ for some n. By construction, $\langle D_n \cup exp\, A_\alpha \rangle$ is dense in $exp\, B_n$, hence

$$0 < y \wedge d \leq x$$

for suitable $y \in exp\, A_\alpha$ and $d \in D_n$. By inductive assumption, $A_\alpha \leq_{reg} A$, so 3.3.14 implies $exp\, A_\alpha \leq_{reg} exp\, A$. This allows us to take $z \in exp\, A_\alpha$ such that

$$(exp\, A_\alpha){\restriction}\,(z - y \wedge d) = \{0\}.$$

As S_α is dense in $exp\, A_\alpha$, we find $s \in S_\alpha$ such that $0 < s \leq z$. Then $0 < s \wedge d$ for, otherwise, $0 < s \in (exp\, A_\alpha){\restriction}\,(z - y \wedge d) = \{0\}$.

Moreover, $s - y \in (exp\, A_\alpha){\restriction}\,(z - y \wedge d) = \{0\}$, i.e. $s \leq y$.

It follows that $0 < s \wedge d \leq y \wedge d \leq x$, and, as $s \wedge d \in T_n \leq S_{\alpha+1}$ the required density is proved.

Immediately from (1) and and the regularity of $\mathcal{S}$, i.e. from

$$S_{\alpha+1} \leq_d exp\, A_{\alpha+1} \quad \text{and} \quad S_{\alpha+1} \leq_{reg} exp\, A$$

we get $exp\, A_{\alpha+1} \leq_{reg} exp\, A$. Indeed, letting $0 < h \in exp\, A$ be given, we choose $0 < s \in S_{\alpha+1}$ such that $S_{\alpha+1}{\restriction}\,(s - h) = \{0\}$. Then s belongs to $exp\, A_{\alpha+1}$ and $exp A_{\alpha+1}{\restriction}\,(s - h) = \{0\}$, by density of $S_{\alpha+1}$.

From $exp\, A_{\alpha+1} \leq_{reg} exp\, A$ we finally get (2), by 3.3.14 again.

The inductive construction is finished and, as explained above, it proves the theorem □

The (easy direction of) the above theorem enables us to determine the exponentials of weakly projective algebras up to co-completeness.

Proposition 5.4.6 *Let A be a weakly projective algebra. If A has no, exactly one, or infinitely many atoms then $A \sim exp\, A$. If A has $1 < n < \omega$ atoms, then $A \times B \sim exp\, A$, where B is a finite Boolean algebra with $2^n - (n+1)$ atoms.*

Proof. Given that $exp\, A$ is also weakly projective, 5.2.5 reduces our task to determining the number of atoms and the π-spectrum of $exp\, A$. As we have to do some calculations in exponentials, we use the notation of section 3.3.

To determine the number of atoms, one stablishes

(1) $v(a_1, \ldots, a_n)$ is an atom of $exp\, A$ iff $a_1, \ldots, a_n$ are atoms of A.

This is almost immediate from the definition of atom and 3.3.3. Given that the element $v(a_1, \ldots, a_n)$ depends only on the set $\{a_1, \ldots, a_n\}$, we conclude that $exp\, A$ has as many atoms as the set of atoms of A has non-empty subsets, i.e. $2^n - 1$, if A has n atoms.

Next we establish the formula

(2) $\pi((exp\, A)\!\restriction v(a_1,\ldots,a_n)) = \max\{\pi(A\!\restriction a_1),\ldots,\pi(A\!\restriction a_n)\}$.

We start by showing $\pi(exp\, A\!\restriction v(a_1,\ldots,a_n)) \geq \pi(A\!\restriction a_1)$, which, by symmetry, yields one of the two necessary inequalities.

Let $C \subseteq exp\, A\!\restriction v(a_1,\ldots,a_n)$ be dense and of power λ. By 3.3.4, we can assume $C = \{v(c_1^\alpha,\ldots,c_{k_\alpha}^\alpha) : \alpha < \lambda\}$. We put $\widehat{C} = \{c_i^\alpha : \alpha < \lambda,\ i \leq k_\alpha\}$ and check that $\widehat{C}$ is dense in $A\!\restriction a_1$. Let $0 \neq d \leq a_1$ be given. Then, as C is dense,

$$0 \neq v(c_1^\alpha,\ldots,c_{k_\alpha}^\alpha) \preceq v(d,a_2,\ldots,a_n)$$

for some α. Assuming $c_i^\alpha - d > 0$ for all i, we get on the one hand

$$0 \neq v(c_1^\alpha - d,\ldots,c_{k_\alpha}^\alpha - d) \preceq v(c_1^\alpha,\ldots,c_{k_\alpha}^\alpha) \preceq v(d,a_2,\ldots,a_n),$$

and, on the other hand,

$$v(c_1^\alpha - d,\ldots,c_{k_\alpha}^\alpha - d) \odot v(d,a_2,\ldots,a_n) = 0,$$

both by 3.3.3. This contradiction proves that $c_i^\alpha - d = 0$ for some i, i.e. $\widehat{C} \ni c_i^\alpha \leq d$. The density of $\widehat{C}$ in $A\!\restriction a_1$ is established and it yields $\lambda = \left|\widehat{C}\right| \geq \pi(A\!\restriction a_1)$.

To prove the other direction of (2), let $C_1 \subseteq A\!\restriction a_1,\ldots,C_n \subseteq A\!\restriction a_n$ be dense. It will be sufficient to show that

$$C = \{v(c_1^1,\ldots,c_{k_1}^1,\ldots,c_1^n,\ldots c_{k_n}^n) : 0 < k_1,\ldots,k_n;\ c_j^i \in C_i\}$$

is dense in $(exp\, A)\!\restriction v(a_1,\ldots,a_n)$. Below each non-zero element of $(exp\, A)\!\restriction v(a_1,\ldots,a_n)$ there is one of the form $v(d_1,\ldots,d_p)$. Let $d_1^i,\ldots d_{k_i}^i$ be the non-zero elements among $a_i \wedge d_1,\ldots,a_i \wedge d_p$. Notice that $k_i > 0$, because, otherwise, $v(\vec{d}) \odot v(\vec{a}) = 0$, by 3.3.3(3). As each d_j^i belongs to $A\!\restriction a_i$, we can take non-zero $c_j^i \in C_i$ such that $c_j^i \leq d_j^i$. Then

$$v(c_1^1,\ldots,c_{k_1}^1,\ldots,c_1^n,\ldots,c_{k_n}^n) \preceq v(d_1^1,\ldots,d_{k_1}^1,\ldots,d_1^n,\ldots,d_{k_n}^n) \preceq v(d_1,\ldots,d_p),$$

as the first of these elements belongs to C, its density is proved and thereby (2).

Some easy considerations now prove that A and $exp\, A$ have the same π-spectrum. We leave them to the reader. $\square$

As exercises along the same line the reader may now enjoy himself proving that for a weakly projective algebra A without or with infinitely many atoms

(1) $A \sim A \times A$,

(2) $A \sim A \otimes A$, and

(3) $A^{(\omega)} \sim F$, where $A^{(\omega)}$ is the free producct of countably many copies of A and F is a free Boolean algebra of cardinality $\pi(A)$.

The above results for exponentials are true for a wider class of functors. Indeed, Shapiro proved

Information 5.4.7 (Theorem 4 in [49]) *If A is weakly projective, then $F(A)$ is also weakly projective, where F can be any normal functor in the sense of Ščepin[5]. The converse is also true* (Shapiro, unpublished). *Moreover, $F(\mathrm{Fr}\,\kappa) \sim \mathrm{Fr}\,\kappa$ for all infinite κ.*

As the functor SP^2 is normal, this result implies, in paticular, that $SP^2(\mathrm{Fr}\,\kappa) \sim \mathrm{Fr}\,\kappa$ for all infinite κ. The reader may find it a nice exercise to prove this in the way we proceeded for exp.

It follows that up to now all our examples of rc-filtered Boolean algebras are (either projective or, at least,) co-complete to projective Boolean algebras. It will cost a whole chapter to prove that this is not so, in general.

5.5 Regularly filtered algebras

In this section we study the class of Boolean algebras that are adequate to regular embeddings. We call them regularly filtered. In many respects they relate to weakly projective algebras as rc-filtered algebras relate to projective ones. It is however an **open problem**

whether each regularly filtered algebra is co-complete to an rc-filtered one.

The converse follows from 5.5.6 below.

Definition 5.5.1 A Boolean algebra is called *regularly filtered* if it has a regular skeleton. □

Let us prove that the class of regularly filtered algebras is adequate with the class of regular embeddings. As axiom (A1) is true, by definition, we only need

Theorem 5.5.2 *If a Boolean algebra has a well-ordered regular filtration consisting of regularly filtered Boolean algebras, then it is regularly filtered itself.*

In the **proof** we follow Ščepin's argument for Theorem 2.3 in [55]. Let κ be a limit ordinal and $(A_\alpha)_{\alpha < \kappa}$ a continuous chain of regular subalgebras of A such that $A = \bigcup_{\alpha < \kappa} A_\alpha$ and each A_α has a regular skeleton S_α, say. We have to construct a regular skeleton of A. The main part of the proof will be an inductive construction of new regular skeletons T_α of A_α, which are, in a sense, compatible with each other.

Before starting we fix mappings $f_\alpha : A \to A_\alpha$ such that $A_\alpha \upharpoonright (f_\alpha(a) - a) = \{0\}$ and $f_\alpha(a) > 0$ if $a > 0$. This is possible because $A_\alpha \leq_{reg} A$.

In the inductive construction we guarantee that the skeletons T_α satisfy the following conditions for all $\alpha < \beta < \kappa$.

(1) $T_\alpha \subseteq T_\beta$

(2) $\{T \cap A_\alpha : T \in T_\beta\} \subseteq T_\alpha$

---

[5] We are not going to define this notion here, just tell the reader that exp and SP^n are normal, while λ is not. Those who really want to know more may consult definition 14 in [55].

(3) For each non-zero $t \in T \in \mathcal{T}_\beta$ there exists a non-zero $t' \in T \cap A_\alpha$ such that $A_\alpha \!\upharpoonright (t' - t) = \{0\}$.

START of the construction. Putting $\mathcal{T}_0 = \mathcal{S}_0$ we have nothing to check.

SUCESSOR STEP. Assume that $\mathcal{T}_\alpha$ has been constructed and put

$$\mathcal{T}_{\alpha+1} = \mathcal{T}_\alpha \cup \left[\, \mathcal{S}_{\alpha+1} \cap \{T \le A_{\alpha+1} : T \cap A_\alpha \in \mathcal{T}_\alpha\} \cap \{T \le A_{\alpha+1} : f_\alpha(T) \subseteq T\} \,\right]$$

Lemmas 1.1.4, 1.1.3, and 1.1.7 show that $\mathcal{T}_{\alpha+1}$ is a skeleton for $A_{\alpha+1}$. It is regular because $\mathcal{S}_{\alpha+1}$ is regular and $\mathcal{T}_\alpha$ is regular in $A_\alpha \le_{reg} A_{\alpha+1}$. The construction explicitly guarantees that conditions (1) and (2) remain true up to $\alpha + 1$.

We check condition (3). Let a non-zero $t \in T \in \mathcal{T}_{\alpha+1}$ and $\delta \le \alpha$ be given. If $T \in \mathcal{T}_\alpha$, then the induction hypothesis works. Otherwise, $0 < f_\alpha(t) \in T \cap A_\alpha$ and $A_\alpha \!\upharpoonright (f_\alpha(t) - t) = \{0\}$. So $f_\alpha(t)$ is the sought for t' if $\delta = \alpha$. If $\delta < \alpha$, we apply the induction hypothesis to $f_\alpha(t) \in T \cap A_\alpha \in \mathcal{T}_\alpha$. It yields $t' \in (T \cap A_\alpha) \cap A_\delta = T \cap A_\delta$ such that $A_\delta \!\upharpoonright (t' - f_\alpha(t)) = \{0\}$. Let us check that this t' works for t, i.e. $A_\delta \!\upharpoonright (t' - t) = \{0\}$.

Indeed, $0 < x \le t'$ and $x \in A_\delta$ imply $0 < x \wedge f_\alpha(t)$, by the choice of t'. From $x \wedge f_\alpha(t) \in A_\alpha$ and $x \wedge f_\alpha(t) \le f_\alpha(t)$ we further get $x \wedge f_\alpha(t) \wedge t > 0$, hence $x \wedge t > 0$, as desired.

LIMIT STEP. Assume that the construction has proceeded up to the limit ordinal λ. We put

$$\mathcal{T}_\lambda = \{T \le A_\lambda : T \cap A_\alpha \in \mathcal{T}_\alpha \text{ for all } \alpha < \lambda\}.$$

First we check that $\mathcal{T}_\lambda$ is a skeleton. Closedness is obvious. To see that $\mathcal{T}_\lambda$ is absorbing, take any $C \le A_\lambda = \bigcup_{\alpha < \lambda} A_\alpha$. By induction on $\alpha \le \lambda$ we find $T_\alpha \in \mathcal{T}_\alpha$ of power $\le |C|$ such that

(o) $C \cap A_0 \le T_0$

(i) $T_\alpha \cup (C \cap A_{\alpha+1}) \subseteq T_{\alpha+1}$

(ii) $T_\gamma = \bigcup_{\alpha < \gamma} T_\alpha$ for limit ordinals.

(iii) If $C \cap A_\alpha = C \cap A_{\alpha+1}$, then $T_\alpha = T_{\alpha+1}$.

Condition (iii) does not contradict $T_{\alpha+1} \in \mathcal{T}_{\alpha+1}$ because $T_\alpha \subseteq T_{\alpha+1}$, by (1). To see that (ii) does not destroy $|T_\gamma| \le |C|$ we evaluate

$$\begin{aligned}
|T_\gamma| &= |T_0| + \sum_{\alpha < \gamma} |T_{\alpha+1} \setminus T_\alpha| \\
&= |T_0| + \sum \{|T_{\alpha+1} \setminus T_\alpha| : \alpha < \gamma \text{ and } C \cap A_\alpha \neq C \cap A_{\alpha+1}\} \\
&\le |C| + (|C| \cdot |C|).
\end{aligned}$$

This shows that the construction is possible. Then T_λ has power $|C|$ and we have $C \le T_\lambda$, by (o) and (i). By (2), each $T_\beta \cap A_\alpha$ belongs to $\mathcal{T}_\alpha$, so

$$T_\lambda \cap A_\alpha = \bigcup_{\alpha \le \beta < \lambda} (T_\beta \cap A_\alpha)$$

is the union of a subchain of T_α, which is closed. It follows that T_λ belongs to T_λ and absorption is proved.

Next we prove that T_λ is regular. Fix $T \in T_\lambda$ and a non-zero $a \in A_\lambda$. We have to find a non-zero $c \in T$ such that $T \upharpoonright (c - a) = \{0\}$. There is some $\alpha < \lambda$ such that $a \in A_\alpha$. As $T \cap A_\alpha \in T_\alpha$ which is regular in A_α, we find $c \in T \cap A_\alpha$ such that $(T \cap A_\alpha) \upharpoonright (c - a) = \{0\}$. We show that c works for T as well. In order to derive a contradiction, we consider any $t \in T$ such that $0 < t \leq c$ but $t \wedge a = 0$.

There is some $\beta < \lambda$ such that $t \in T \cap A_\beta \in T_\beta$. There is no harm in assuming $\alpha < \beta$. Thus, by (3), $A_\alpha \upharpoonright (t' - t) = \{0\}$ for some $0 < t' \in T \cap A_\alpha$. Then $t' \wedge c \wedge a \in A_\alpha \upharpoonright (c - a) = \{0\}$, hence $0 < t' \wedge c \in A_\alpha \upharpoonright (c - a) = \{0\}$, which is the desired contradiction.

Finally, we have to check that conditions (1) - (3) are preserved, provided they are true below λ.

(1) $T_\alpha \subseteq T_\lambda$: For $T \in T_\alpha$ and $\beta < \lambda$ we have, by (1) and (2), $T \cap A_\beta \in T_\beta$ for all $\beta < \lambda$. So, $T \in T_\lambda$.

(2) is explicitly constructed.

(3) Let a non-zero $t \in T \in T_\lambda$ and some $\alpha < \lambda$ be given. Fix $\alpha < \beta < \lambda$ such that t belongs to $T \cap A_\beta \in T_\beta$. By induction hypothesis, i.e. (3) for $\alpha < \beta$, there exists $t' \in (T \cap A_\beta) \cap A_\alpha = T \cap A_\alpha$ such that $A_\alpha \upharpoonright (t' - t) = \{0\}$, as desired. The inductive construction is finished.

It remains to put $T = \{T \leq A : T \cap A_\alpha \in T_\alpha$ for all $\alpha < \kappa\}$ The same argument as in the limit step shows that T is a regular skeleton of A. The theorem is proved. $\square$

The next result is analogous to 2.2.7 and proved in the same way.

Proposition 5.5.3 *Assume that A is regularly filtered and let $M \subseteq A$ have power $\leq \aleph_1$. Then there exists a weakly projective algebra B such that $M \subseteq B \leq_{reg} A$. In particular,*

(1) *A has the ccc, and*

(2) *if $\pi(A) \leq \aleph_1$, then A is weakly projective.*

Proof. The assertion is an immediate consequence of 1.1.6 and 5.3.1. The ccc then follows from 5.1.5. $\square$

Remark. The above proof of the ccc is a bit heavy because of the involvement of 5.3.1. An alternative approach uses 1.1.6 to reduce it to the assertion:

If A has a regular well-ordered filtration $(A_\alpha)_{\alpha < \omega_1}$ consisting of countable subalgebras, then A satisfies the ccc.

This is the Solovay-Tennenbaum Lemma (Theorem 6.3 in [58]), popular in forcing. Alternatively, the reader can readapt our proof of 2.10.2. $\square$

We now extend the above theorem on chains to arbitrary σ-complete filtrations.

Theorem 5.5.4 *If A has a σ-complete regular filtration all elements of which are regularly filtered, then A is regularly filtered, too.*

Proof. Let $(A_i)_{i \in I}$ be the regular filtration of A in question.

CLAIM 1. *A satisfies the ccc.*

Indeed, let $M \subseteq A$ have power $\aleph_1$. By 1.1.6, we find a continuous chain of indices such that $M \subseteq \bigcup_{\alpha < \omega_1} A_{i_\alpha}$. The union is regularly filtered, by 5.5.2, and, therefore, satisfies the ccc, by 5.5.3. So M does not consist of pairwise disjoint elements.

CLAIM 2. *If $J \subseteq I$ is directed, then $A_J = \bigcup_{j \in J} A_j$ is a regular subalgebra of A.*

We use the test 1.2.16(4). Let $M \subseteq A_J$ be maximal disjoint. We have to verify that M is also maximal in A. By claim 1, we can enumerate M as $\{m_n : n < \omega\}$. Directedness of J then allows us to pick inductively $j_0 \leq j_1 \leq j_2 \leq \ldots$ in J such that $m_n \in A_{j_n}$. By σ-closedness, $i = \sup\{j_n : n < \omega\}$ exists in I. Although i does not necessarily belong to J, we have $A_i = \bigcup_{n < \omega} A_{j_n} \leq A_J$. Being maximal even in A_J, $M \subseteq A_i$ is maximal in A_i. As $A_i \leq_{reg} A$, M is also maximal in A. The second claim is also proved. It reduces the result to Iwamura's Lemma and 5.5.2 exactly as in the proof of 2.1.5. $\square$

Corollary 5.5.5 *The following assertions are equivalent for an arbitrary Boolean algebra A.*

(1) *A is regularly filtered.*

(2) *$\{C \leq_{reg} A : |C| \leq \aleph_0\}$ contains a club.*

(3) *A has a σ-complete regular filtration consisting of countable subalgebras.*

Proof. Indeed, the countable members of the regular skeleton form the club. The club indexed by itself is the filtration. As all countable Boolean algebras are trivially regularly filtered, the above theorem yields the last implication. $\square$

Next we discuss some more closure properties of the class of regularly filtered Boolean algebras.

Proposition 5.5.6 *If A is co-complete to a regularly filtered Boolean algebra, then A is itself regularly filtered. In particular, Boolean algebras that are co-complete to rc-filtered ones, are regularly filtered.*

Proof. By 5.1.4, this result follows from the two special cases below.

Lemma 5.5.7 *If $A \leq_d B$ and A has a regular skeleton, then B also has a regular skeleton.*

Proof. Fix a mapping $f : B \to A$ such that $f(0) = 0$, and $0 < f(b) \leq b$ for all non-zero $b \in B$. Let S be a regular skeleton of A. We know (1.1.4) that $T = \{T \leq B : T \cap A \in S\}$ is a skeleton of B. Intersecting it with the fixed-point skeleton of f we get the desired regular skeleton of B, for

CLAIM $T \cap A \leq_{reg} A$ *and* $f(T) \subseteq T$ *imply* $T \leq_{reg} B$.

To see this. let $b \in B \setminus \{0\}$ be given. Choose $a \in A \setminus \{0\}$ such that $a \leq b$. As $T \cap A \leq_{reg} A$, there exists a non-zero $t \in T \cap A$ such that

$$(T \cap A){\restriction}(t - a) = \{0\}.$$

We show that $T{\restriction}(t-b) = \{0\}$. Consider any $s \in T$ such that $s \leq t$ and $s \wedge b = 0$. Then $f(s) \leq s \leq t$ and $f(s) \wedge a \leq s \wedge b = 0$, i.e. $f(s) \in (A \cap T){\restriction}(t - a) = \{0\}$. But from $f(s) = 0$, we get $s = 0$, by the definition of f. □

As dense embeddings are regular, the next result yields the nessecary second special case of 5.5.6. It is the counterpart of 2.3.1(1) in the realm of regularly filtered Boolean algebras.

Proposition 5.5.8 *If* $A \leq_{reg} B$ *and* B *has a regular skeleton, then* A *also has a regular skeleton.*

Proof. Fix a mapping $f : B \rightarrow A$ such that

(1) $0 < f(b)$ and $A{\restriction}(f(b) - b) = \{0\}$

for all non-zero $b \in B$. Let $\mathcal{S}$ be a regular skeleton of B. Intersecting it with the fixed-point skeleton of f we can assume that all $S \in \mathcal{S}$ are closed under f. Given that $(S \cap A)_{S \in \mathcal{S}}$ is a σ-complete filtration of A, it is sufficient to verify $S \cap A \leq_{reg} A$ for all $S \in \mathcal{S}$.

     Let $a \in A \setminus \{0\}$ be given. Take $s \in S \setminus \{0\}$ such that

(2) $S{\restriction}(s - a) = \{0\}$.

Then $0 < f(s) \in A \cap S$ and we claim that $(A \cap S){\restriction}(f(s) - a) = \{0\}$.

     Indeed, each $t \in A \cap S$ such that $t \leq f(s)$ and $t \wedge a = 0$ satisfies $s \wedge t \in S$, $s \wedge t \leq s$ and $s \wedge t \wedge a = 0$, i.e. $s \wedge t \in S{\restriction}(s - a) = \{0\}$, by (2). But then $t \in A{\restriction}(f(s) - s) = \{0\}$, by (1). □

From 5.5.6, which is now proved, it should be clear that there are complete regularly filtered Boolean algebras. This shows that in contrast to the rc-filtered case, regularly filtered Boolean algebras may contain uncountable chains. Uncountable well-ordered chains are however still excluded by the ccc. The Bockstein separation property which was quite important for rc-filtered Boolean algebras can also fail as Engelking's example 2.3.4 shows.

The final result in this section is the analog of Ivanov's theorem 3.2.6 in the realm of regularly filtered algebras.

Theorem 5.5.9 *For a Boolean algebra* A *the following assertions are equivalent.*

     (1) A *is regularly filtered.* (2) λA *is weakly projective.* (3) λA *is regularly filtered.*

Proof. Recall that the functor λ reflects and strongly preserves regularity (3.2.8). (1) $\Rightarrow$ (2). Replacing $\leq_{rc}$ by $\leq_{reg}$ in the proof of 3.1.2 and referring to 5.5.7 instead of 2.3.1, we get the following

FACT. *Let F be a continuous functor that preserves cardinalities and strongly preserves regularity of embeddings. If A has a skeleton consisting of regular subalgebras, then $F(A)$ has a regular filtration $(C_\alpha)_{\alpha<\rho}$ starting with a countable C_0 such that each $C_{\alpha+1}$ is countably generated over C_α.*

As λ has all the required properties, the fact yields a representation of λA, which witnesses the co-completeness with a projective algebra, by 5.3.1.

(2) $\Rightarrow$ (3) follows from 5.5.6.

(3) $\Rightarrow$ (1) follows from the reflection of regularity and 3.1.1(4). $\square$

Chapter 6

The twisted embedding and its applications

In this chapter we construct more examples of rc-filtered non-projective Boolean algebras. The first example (in section 3) is not even co-complete with a projective Boolean algebra.

In section 4 we present a construction due to Fuchino which yields $2^{\aleph_2}$ pairwise non-isomorphic rc-filtered Boolean algebras of power $\aleph_2$. If one assumes $V = L$, then the maximal number is attained for all cardinals that are $\geq \aleph_2$ and not weakly compact. Notice that (at least consistently) most of these algebras are not projective, for Koppelberg proved that there are only $2^{<\kappa}$ projective Boolean algebras of cardinality κ if κ is regular and uncountable (Theorem 4.11 in [35]).

All the examples above arise from a peculiar embedding of $\operatorname{Fr}\omega_1$ into itself. It is due to the second author and will be constructed in section 2. Section 1 prepares the construction by providing the basic building block.

The whole construction is due to the second author. The Boolean algebraic translation below is the result of the joint efforts[1] of Sabine Koppelberg, Sakaé Fuchino and the first author.

The material of section 1.4 is needed in this chapter, and the notation $A \leq_{free} B$, $A \leq_s B$, and $A \leq_{sp} B$ occuring below was introduced there.

6.1 The basic square

The aim of this section is to construct what will be called the *basic square*. It consists of countable atomless Boolean algebras

---

[1] It became possible after Ingo Bandlow explained the topological version in great detail at a seminar in Berlin.

$$A' \leq B'$$
$$\text{VI} \qquad \text{VI}$$
$$A \leq B$$

such that the following conditions are satisfied.

(1b) There is an atomless subalgebra $X \leq A'$ which is independent of B and such that $A' = \langle A \cup X \rangle$.

(2b) $A \leq_{free} B$

(3b) $B \leq_{free} B'$

(4b) $A' \leq_{free} B'$

(5b) $\langle A' \cup B \rangle \leq_s B'$, i.e., every ultrafilter of $\langle A' \cup B \rangle$ extends to at least two different ultrafilters of B'.

(6b) $\langle A' \cup B \rangle$ is not regular in B'.

(7b) If $0 < b' \in B'$, then $\langle A' \cup B \rangle$ is not dense in $B' \restriction b'$.

First we give a topological sketch of the construction. Let C denote the Cantor space 2^ω. It is clear that the algebras $Clop\,(C^n)$ are countable and atomless for all $n > 0$, as are the algebras[2] $Clop\,(C^n \oplus C^m)$ for $m, n > 0$. The basic square will be the dual of the outer quadrilateral of the following commutative diagram of continuous surjective mappings.

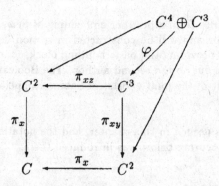

The mappings π are projections, as indicated (e.g. $\pi_{xz} : (x, y, z) \mapsto (x, z)$). The mapping φ splits into $\pi_{xyz} \oplus \psi$, where $\psi : C^3 \to C^3$ is the projection onto the diagonal: $(x, y, z) \mapsto (x, y, y)$.

The space C^3 in the centre of the diagram is not an explicit part of the basic square, but it represents the Boolean algebra $\langle A' \cup B \rangle$, which features in properties (5b) $-$ (7b). The additional 'dimension' of C^4 is responsible for (7b), whereas

----

[2]Here $\oplus$ denotes the disjoint union of topological spaces.

the fact that the diagonal $\psi(C^3)$ is nowhere dense in C^3 implies ($6b$). Notice also that $\pi_{xy} \circ \psi : (x, y, z) \mapsto (x, y)$ and $\pi_{xz} \circ \psi : (x, y, z) \mapsto (x, y)$ are both open mappings with perfect fibres, and this yields ($3b$) and ($4b$). Conditions ($1b$) and ($2b$) are obvious.

For those readers who feel more secure with formulas and calculations we use the rest of this section to give the same description of the basic square in more detail using algebraic language. Let F denote a countable atomless (= countable free) Boolean algebra. We put

$$B' = F \otimes F \otimes F \otimes F \times F \otimes F \otimes F$$

and

$$A = \{(a \otimes 1 \otimes 1 \otimes 1, a \otimes 1 \otimes 1) : a \in F\} \leq B'.$$

We further let A' be the set of all finite unions of elements of

$$\{(a \otimes b \otimes 1 \otimes 1, a \otimes b \otimes 1) : a, b \in F\}$$

and B the set of all finite unions of elements of

$$\{(a \otimes 1 \otimes c \otimes 1, a \otimes c \otimes 1) : a, c \in F\}.$$

Checking that A' and B are indeed subalgebras of B' requires some straightforward calculations. Alternatively, one can consider the homomorphisms $\varphi_{1/2/3} : F \to B'$ sending $x \in F$ to

$$(x \otimes 1 \otimes 1 \otimes 1, x \otimes 1 \otimes 1), \quad (1 \otimes x \otimes 1 \otimes 1, 1 \otimes x \otimes 1), \quad (1 \otimes 1 \otimes x \otimes 1, 1 \otimes x \otimes 1)$$

respectively. Then A' is the image of $F \otimes F$ under $\varphi_1 \otimes \varphi_2$ and B is the image of $F \otimes F$ under $\varphi_1 \otimes \varphi_3$.

In the rest of this section we check that the algebras described above have properties ($1b$) − ($7b$).

The subalgebra X demanded in ($1b$) is of course

$$X = \{(1 \otimes x \otimes 1 \otimes 1, 1 \otimes x \otimes 1) : x \in F\}.$$

Analogously, the subalgebra

$$Y = \{(1 \otimes 1 \otimes y \otimes 1, 1 \otimes y \otimes 1) : y \in F\}$$

of B witnesses $A \leq_{free} B$, i.e. ($2b$).

Condition ($3b$), i.e., $B \leq_{free} B'$ will be established with the help of Sirota's Lemma 1.4.10 reducing it to two claims

$$(i) \quad B \leq_{rc} B' \qquad and \qquad (ii) \quad B \leq_{sp} B'.$$

To prove (i) we use 1.2.3. Each element of B' is a finite union of elements of one of the forms

$$\beta_1 = (a \otimes b \otimes c \otimes d, \, 0) \quad \text{or} \quad \beta_2 = (0, \, a \otimes b \otimes c).$$

We check that

$$\alpha_1 = (a \otimes 1 \otimes c \otimes 1, \, a \otimes c \otimes 1) \quad \text{and} \quad \alpha_2 = (a \otimes 1 \otimes b \otimes 1, \, a \otimes b \otimes 1)$$

are minimal elements of B above β_1 and β_2, respectively.

If this were not the case for α_1, then we could take some $\gamma \in B$ such that $\gamma \wedge \beta_1 = 0$ but $\gamma \wedge \alpha_1 \neq 0$. Without loss, γ is of the form

$$\gamma = (a' \otimes 1 \otimes c' \otimes 1, \, a' \otimes c' \otimes 1)$$

with $a', c' \in F$. From

$$\gamma \wedge \beta_1 = (a' \wedge a \otimes b \otimes c' \wedge c \otimes 1, \, 0) = 0$$

we get $a' \wedge a = 0$ or $c' \wedge c = 0$. In both cases it follows that

$$\gamma \wedge \alpha_1 = (a' \wedge a \otimes b \otimes c' \wedge c \otimes 1, \, a' \wedge a \otimes c' \wedge c \otimes 1) = 0,$$

a contradiction. The same argument works for $\beta_2 \leq \alpha_2$.

To prove (ii), we let E denote the subalgebra of B' consisting of all finite unions of elements of

$$\{(a \otimes b \otimes c \otimes 1, \, d \otimes e \otimes 1) : a, b, c, d, e \in F\}.$$

The subalgebra

$$Z = \{(1 \otimes 1 \otimes 1 \otimes x, \, 1 \otimes 1 \otimes y) : x, y \in F\}$$

witnesses $E \leq_{free} B'$, hence $E \leq_{sp} B'$, by 1.4.8. From $B \leq E$ we then get $B \leq_{sp} B'$, by 1.4.7. Thus condition $(3b)$ is proved and in the same way one proves $(4b)$.

To establish $(5b)$, we use the same subalgebra E as above. Noticing that $A' \leq E$, we get $\langle A' \cup B \rangle \leq E \leq_{sp} B'$. Again by 1.4.7, we have $\langle A' \cup B \rangle \leq_{sp} B'$, which is more than we needed to show.

To prove $(6b)$, i.e., the non-regularity of $\langle A' \cup B \rangle$ in B', we consider the element $\varepsilon = (0, 1 \otimes 1 \otimes 1) \in B'$ and assume that there were some $0 < \delta \in \langle A' \cup B \rangle$ such that $\langle A' \cup B \rangle \restriction (\delta - \varepsilon) = \{0\}$. Without loss we can assume that δ is of the form

$$\delta = (a \otimes b \otimes c \otimes 1, \, a \otimes b \wedge c \otimes 1).$$

Indeed, $\langle A' \cup B \rangle$ consists of all elements of the form $\bigvee_{i=1}^{n} \alpha'_i \wedge \beta_i$, with $\alpha'_i \in A'$ and $\beta_i \in B$. The elements α'_i and β_i are themselves finite unions of elements of the forms

$$(a \otimes b \otimes 1 \otimes 1, \, a \otimes b \otimes 1) \quad \text{and} \quad (a \otimes 1 \otimes c \otimes 1, \, a \otimes c \otimes 1),$$

respectively. By distributivity and shrinking to one disjunctive term, we get the desired form for δ.

From $\delta \neq 0$ we must have $b \wedge c \neq 0$. As F is atomless, we can find non-zero $b', c' \leq b \wedge c$ such that $b' \wedge c' = 0$. Then

$$0 < \gamma = (a \otimes b' \otimes c' \otimes 1, \, a \otimes b' \wedge c' \otimes 1) = (a \otimes b' \otimes c' \otimes 1, 0)$$

belongs to $\langle A' \cup B \rangle \restriction (\delta - \varepsilon)$, and this is a contradiction.

Finally, we check (7b). Take an arbitrary non-zero $\beta \in B'$. We find a non-zero $\gamma \in B'$ such that $\langle A' \cup B \rangle \restriction \gamma = \{0\}$ and even $E \restriction \gamma = \{0\}$, for the E introduced above.

Making β smaller if necessary, we can assume it to have the form

$$\beta = (a \otimes b \otimes c \otimes d, \, e \otimes f \otimes g).$$

From $0 < \beta$ we get $0 < d$. As F is atomless we may take $0 < d' < d$. In particular, $d' \neq 1$, hence

$$0 < \gamma = (a \otimes b \otimes c \otimes d', \, e \otimes f \otimes g) \leq \beta$$

and there is no element of $E \supseteq \langle A' \cup B \rangle$ below γ.

All of the conditions have now been checked and so the construction of the basic square is complete.

6.2 Twisted embeddings

For want of a better name we introduce the following

Definition 6.2.1 An embedding (or extension) $A \leq B$ will be called *twisted* iff it has the following properties.

(1) A and B are infinite and free.

(2) $A \leq_{rc} B$

(3) There is no countable subset $W \subseteq B$ such that $\langle A \cup W \rangle$ is dense in some $B \restriction b$ for $0 < b \in B$.

(4) There is a filtration $(B_\alpha)_{\alpha < \omega_1}$ of B such that for all $\alpha < \omega_1$

    (4i) $A \leq B_\alpha$ and $w(B_\alpha / A) = \aleph_0$, and

    (4ii) B_α is not a regular subalgebra of B. $\square$

Notice that conditions $(1) - (3)$ and $(4i)$ are also true of the standard embedding $\operatorname{Fr} \kappa \leq \operatorname{Fr} \kappa \otimes \operatorname{Fr} \omega_1$. So the twistedness is concentrated in condition $(4ii)$. Notice also that (3) and $(4i)$ imply $w(B/A) = \aleph_1$ for twisted embeddings.

Observation 6.2.2 *If $A \leq B$ is twisted and $(C_\alpha)_{\alpha<\omega_1}$ is any filtration of B such that $A \leq C_\alpha$ and $w(C_\alpha/A) = \aleph_0$ for all α, then $\{\alpha < \omega_1 : C_\alpha \leq_{reg} B\}$ is not closed and unbounded.*

Indeed, the usual argument shows that $C_\alpha = B_\alpha$ for a closed and unbounded set of indices. $\square$

Next we prove that twisted embeddings exist. The idea is to iterate the basic square ω_1 times. More precisely, we construct a chain $(A_\alpha \leq B_\alpha)_{\alpha \leq \omega_1}$ of embeddings such that

$$
\begin{array}{ccc}
A_\beta & \leq & B_\beta \\
\text{VI} & & \text{VI} \\
A_\alpha & \leq & B_\alpha
\end{array}
$$

and the following conditions are satisfied for all $\alpha < \beta \leq \omega_1$.

(0c) If $\beta < \omega_1$, then $|B_\beta| = \aleph_0$.

(1c) There exists a free subalgebra $X_{\alpha\beta} \leq A_\beta$ which is independent of B_α and satisfies $A_\beta = \langle A_\alpha \cup X_{\alpha\beta} \rangle$.

(2c) $B_\alpha \leq_{free} B_\beta$.

(3c) $A_\beta \leq_{rc} B_\beta$.

(4c) $A_\beta \leq_{sp} B_\beta$.

(5c) $\langle A_\beta \cup B_\alpha \rangle \leq_s B_\beta$.

(6c) $\langle A_\beta \cup B_\alpha \rangle$ is not regular in B_β.

(7c) If $0 < b \in B_\beta$, then $\langle A_\beta \cup B_\alpha \rangle$ is not dense in $B_\beta \restriction b$.

START. Choose countable atomless $A_0 \leq B_0$ such that $A_0 \leq_{free} B_0$.

SUCCESSOR STEP. Assume that $A_\alpha \leq B_\alpha$ have been constructed for all $\alpha \leq \gamma$ such that all conditions hold for $\alpha < \beta \leq \gamma$. In particular, by Sirota's Lemma 1.4.10, it follows from (0c), (3c), and (4c) that $A_\gamma \leq_{free} B_\gamma$. Consequently, we may take Boolean algebras $A_{\gamma+1}$ and $B_{\gamma+1}$ such that

$$
\begin{array}{ccc}
A_{\gamma+1} & \leq & B_{\gamma+1} \\
\text{VI} & & \text{VI} \\
A_\gamma & \leq & B_\gamma
\end{array}
$$

is isomorphic to the basic square. To check that the conditions remain true up to $\gamma+1$, it is sufficient to consider $\alpha < \beta = \gamma+1$. If $\alpha = \gamma$, then conditions (0c) – (7c) follow from conditions (1b)–(7b) for the basic square.

Assume then that $\alpha < \gamma$. By (1b), there exists a free subalgebra $X \leq A_{\gamma+1}$ which is independent of B_γ and satisfies $A_{\gamma+1} = \langle A_\gamma \cup X \rangle$. By inductive assumption, $A_\gamma = \langle A_\alpha \cup X_{\alpha\gamma} \rangle$ for some free $X_{\alpha\gamma} \leq A_\gamma$ which is independent of B_α. We put $X_{\alpha\,\gamma+1} = \langle X_{\alpha\gamma} \cup X \rangle$. Then

$$A_{\gamma+1} = \langle A_\gamma \cup X \rangle = \langle \langle A_\alpha \cup X_{\alpha\gamma} \rangle \cup X \rangle = \langle A_\alpha \cup \langle X_{\alpha\gamma} \cup X \rangle \rangle = \langle A_\alpha \cup X_{\alpha\,\gamma+1} \rangle.$$

To see that $X_{\alpha\,\gamma+1}$ is free, it is sufficient to establish that it is atomless. But any hypothetical atom must be of the form $y \wedge z$ with $y \in X_{\alpha\gamma}$ and $z \in X$. As X is free, z can be split: $z = z_1 \vee z_2$. Hence $y \wedge z = (y \wedge z_1) \vee (y \wedge z_2)$ and both meets are non-empty, by independence of X and $X_{\alpha\gamma} \subseteq B_\gamma$.

To see that $X_{\alpha\,\gamma+1}$ is also independent of B_α, we take any $0 < b \in B_\alpha$ and $x \in X_{\alpha\,\gamma+1}$. Write $x = \bigvee_{i=1}^n y_i \wedge z_i$, where $y_i \in X_{\alpha\gamma}$ and $z_i \in X$. It suffices to prove $b \wedge y_1 \wedge z_1 \neq 0$. But from $b \wedge y_1 \wedge z_1 = 0$ and the independence of $B_\gamma \supseteq \langle B_\alpha \cup X_{\alpha\gamma} \rangle \ni b \wedge y_1$ from $X \ni z_1$ we would get $b \wedge y_1 = 0$, which is impossible since B_α and $X_{\alpha\gamma}$ are independent. Thus condition (1c) is proved.

By transitivity, $B_\alpha \leq_{free} B_{\gamma+1}$ follows from $B_\alpha \leq_{free} B_\gamma$ (induction hypothesis) and $B_\gamma \leq_{free} B_{\gamma+1}$ (condition (3b)).

Conditions (3c) and (4c) follow from (4b). Conditions (5c) and (6c) deal with the embedding

$$\langle A_{\gamma+1} \cup B_\alpha \rangle \leq B_{\gamma+1}.$$

By the induction hypothesis, the corrsponding conditions hold for

$$\langle A_{\alpha+1} \cup B_\alpha \rangle \leq B_{\alpha+1}.$$

To lift them, we apply assertions (1) and (2) of the following lemma to the diagram

$$\begin{array}{ccc} \langle A_{\gamma+1} \cup B_\alpha \rangle & \leq & B_{\gamma+1} \\ \mathsf{VI} & & \mathsf{VI} \\ \langle A_{\alpha+1} \cup B_\alpha \rangle & \leq & B_{\alpha+1} \end{array}$$

which is possible, since $X_{\alpha+1\,\gamma+1}$ is independent of $B_{\alpha+1} \geq \langle A_{\alpha+1} \cup B_\alpha \rangle$, and since

$$\langle A_{\gamma+1} \cup B_\alpha \rangle = \langle \langle A_{\alpha+1} \cup X_{\alpha+1\,\gamma+1} \rangle \cup B_\alpha \rangle = \langle \langle A_{\alpha+1} \cup B_\alpha \rangle \cup X_{\alpha+1\,\gamma+1} \rangle.$$

Assertion (3) will be needed in the limit step below.

Lemma 6.2.3 *Consider four Boolean algebras*

$$\begin{array}{ccc} A' & \leq & B' \\ \mathsf{VI} & & \mathsf{VI} \\ A & \leq & B \end{array}$$

and assume that A' has a free subalgebra X which is independent of B and such that $A' = \langle A \cup X \rangle$.

(1) *If $A \leq_s B$, then $A' \leq_s B'$.*

(2) *If $A' \leq_{reg} B'$, then $A \leq_{reg} B$.*

(3) *If $0 < b \in B$ and A is not dense in $B\restriction b$, then A' is not dense in $B\restriction b$ either, let alone in $B'\restriction b$.*

Proof.   (1) Let U' be an ultrafilter of A'. From $A \leq_s B$ we get ultrafilters $V_1 \neq V_2$ of B such that $U' \cap A = V_1 \cap A = V_2 \cap A$. It is sufficient to verify that $U' \cup V_1$, say, extends to an ultrafilter of B'. If this is not true, we can find $a' \in U'$ and $b \in V_1$ such that $a' \wedge b = 0$. Write $a' = \bigvee_{i=1}^{n} a_i \wedge x_i$ with $a_i \in A$ and $x_i \in X$. Assume without loss that $a_1 \wedge x_1 \in U'$.

Then $0 = a_1 \wedge x_1 \wedge b = (a_1 \wedge b) \wedge x_1$ yields $a_1 \wedge b = 0$, by independence of $B \ni a_1 \wedge b$ and X. On the other hand, $a_1 \geq a_1 \wedge x_1 \in U'$, hence $a_1 \in U' \cap A \subseteq V_1 \ni b$ and this is a contradiction.

(2) As X witnesses $A \leq_{free} A'$, we know that $A \leq_{rc} A'$, hence $A \leq_{reg} A'$. Together with $A' \leq_{reg} B'$ this yields $A \leq_{reg} B'$, hence $A \leq_{reg} B$.

(3) Assume that A' is dense in $B\restriction b$ and let some $d \in B$ with $0 < d \leq b$ be given. We have to find $0 < a \in A$ such that $a \leq d$. There is some $A' \ni a' = \bigvee_{i=1}^{n} a_i \wedge x_i \leq d$ (where $a_i \in A$, $x_i \in X$, and $a_i \wedge x_i \neq 0$). Then $a_1 \wedge x_1 \leq d$, hence $(a_1 - d) \wedge x_1 = 0$. By independence, $a_1 - d = 0$, i.e., $a_1 \leq d$ as desired. $\square$

As explained above, the lemma gives us (5c) and (6c). Condition (7c) says that $\langle A_{\gamma+1} \cup B_\alpha \rangle$ is not dense in any $B_{\gamma+1}\restriction b$. But this is obvious, since, by (7b) and construction, not even $\langle A_{\gamma+1} \cup B_\gamma \rangle$ is dense in $B_{\gamma+1}\restriction b$, unless $b = 0$. This ends the successor step.

In the LIMIT STEP we have no choice. To make the chain continuous, we must put $A_\lambda = \bigcup_{\alpha < \lambda} A_\alpha$ and $B_\lambda = \bigcup_{\alpha < \lambda} B_\alpha$. Again we go through the conditions. If $\alpha < \beta < \lambda$, then all conditions are satisfied, by inductive assumption. We must check them for $\alpha < \beta = \lambda$. As long as $\lambda < \omega_1$, everything is countable, so $|B_\lambda| = \aleph_0$, i.e. (0c) is true.

For each $\alpha \leq \gamma < \lambda$ there exist free $X_{\gamma\,\gamma+1} \leq A_{\gamma+1}$ independent of B_γ and such that $A_{\gamma+1} = \langle A_\gamma \cup X_{\gamma\,\gamma+1} \rangle$. We put

$$X_{\alpha\,\lambda} = \langle \bigcup_{\alpha \leq \gamma < \lambda} X_{\gamma\,\gamma+1} \rangle.$$

By induction on $\gamma < \lambda$, one easily proves that

$$A_\gamma \leq \langle A_\alpha \cup X_{\alpha\,\lambda} \rangle \leq A_\lambda$$

hence $A_\lambda = \bigcup_{\gamma < \lambda} A_\gamma \leq \langle A_\alpha \cup X_{\alpha\,\lambda} \rangle$.

To check that $X_{\alpha\,\lambda}$ is independent of B_α, we have to show $b \wedge x \neq 0$ for all non-zero $b \in B_\alpha$ and $x \in X_{\alpha\,\lambda}$. Without loss (in general, x is a finite union of such elements) we can assume $x = x_1 \wedge x_2 \wedge \ldots \wedge x_n$, where $x_i \in X_{\gamma_i\,\gamma_i+1}$ with $\gamma_1 < \gamma_2 < \ldots \gamma_n$.

By induction on $0 \leq k \leq n$ we check that

$$b \wedge x_1 \wedge x_2 \wedge \ldots \wedge x_k \neq 0.$$

This is obvious for $k = 0$. In the induction step we first notice that

$$b \wedge x_1 \wedge x_2 \wedge \ldots \wedge x_k \in \langle B_\alpha \cup X_{\gamma_1 \, \gamma_1 + 1} \cup X_{\gamma_2 \, \gamma_2 + 1} \cup \ldots \cup X_{\gamma_k \, \gamma_k + 1} \rangle \leq B_{\gamma_k + 1} \leq B_{\gamma_{k+1}}.$$

As $x_{k+1} \in X_{\gamma_{k+1} \, \gamma_{k+1} + 1}$, which is independent of $B_{\gamma_{k+1}}$, we have

$$(b \wedge x_1 \wedge x_2 \wedge \ldots \wedge x_k) \wedge x_{k+1} \neq 0,$$

as desired. In a similar way (putting $b = 1$) one verifies that $X_{\alpha \lambda}$ is free. We have established condition (1c).

Condition (2c) follows from transitivity again. To prove (3c), i.e., $A_\lambda \leq_{rc} B_\lambda$, we consider any $b \in B_\lambda$ and find a maximal element of A_λ below b. For some $\gamma < \lambda$ we have $b \in B_\gamma$. By assumption, $A_\gamma \leq_{rc} B_\gamma$. Let a be maximal in A_γ below b. We show that the same a is maximal in the whole of A_λ.

Consider any $a' \in A_\lambda = \langle A_\gamma \cup X_{\gamma \lambda} \rangle$ such that $a' \leq b$. Write $a' = \bigvee_{i=1}^{n} a_i \wedge x_i$, in the obvious way. Then $a' \leq b$ implies $(a_i - b) \wedge x_i = 0$. By independence, $a_i - b = 0$, i.e., $a_i \leq b$. The maximality of a then yields $a_i \leq a$, for all i, hence $a' \leq a$, as claimed.

Condition (4c) says $A_\lambda \leq_{sp} B_\lambda$ and will clearly follow once we know that $\langle A_\lambda \cup B_\alpha \rangle \leq_s B_\lambda$ for all $\alpha < \lambda$. In other words, condition (4c) is weaker than (5c), which we prove next together with (6c). They follow in the same way as in the successor step from the above lemma applied to the diagram

$$\langle A_\lambda \cup B_\alpha \rangle = \langle (A_{\alpha+1} \cup B_\alpha) \cup X_{\alpha+1 \, \lambda} \rangle \leq B_\lambda$$
$$\vee| \qquad\qquad\qquad\qquad \vee|$$
$$\langle A_{\alpha+1} \cup B_\alpha \rangle \qquad\qquad \leq B_{\alpha+1}.$$

Condition (7c) follows from assertion (3) of the same lemma. Indeed, each $0 < b \in B_\lambda$ belongs to some B_γ with $\gamma \geq \alpha$. As $\langle A_{\gamma+1} \cup B_\gamma \rangle$ is not dense in $B_{\gamma+1} \upharpoonright b$, by hypothesis, the lemma (applied to the appropriate diagram) shows that $\langle A_\lambda \cup B_\gamma \rangle$ is not dense in $B_\lambda \upharpoonright b$. It follows that $\langle A_\lambda \cup B_\alpha \rangle$ is not dense either.

The construction of the sequence is complete. It yields the following

Proposition 6.2.4 *There is a twisted embedding $A \leq B$ with $A \simeq \mathrm{Fr}\,\omega_1 \simeq B$. More precisely, for the standard filtration $(A_\alpha)_{\alpha < \omega_1}$ of $A \simeq \mathrm{Fr}\,\omega_1$ there exists a twisted embedding $A \leq B \simeq \mathrm{Fr}\,\omega_1$ such that $A_\alpha \leq_{free} B$ for all α.*

Proof. The embedding $A_{\omega_1} \leq B_{\omega_1}$ is as desired.

(1) $A_{\omega_1} \simeq \mathrm{Fr}\,\omega_1 \simeq B_{\omega_1}$

Indeed, A_0 and B_0 are free and $A_0 \leq_{free} A_{\omega_1}$ and $B_0 \leq_{free} B_{\omega_1}$, by (1c) and (2c), respectively.

(2) $A_{\omega_1} \leq_{rc} B_{\omega_1}$

This is $(3c)$ for $\beta = \omega_1$.

(3) There is no countable subset $W \subseteq B_{\omega_1}$ such that $\langle A_{\omega_1} \cup W \rangle$ is dense in some $B_{\omega_1} \restriction b$ with $0 < b \in B_{\omega_1}$.

As any countable W is contained in some B_α with $\alpha < \omega_1$, this follows immediately from $(7c)$ with $\beta = \omega_1$.

(4) The filtration $((A_{\omega_1} \cup B_\alpha))_{\alpha < \omega_1}$ of B_{ω_1} satisfies for all $\alpha < \omega_1$

(4i) $A_{\omega_1} \leq \langle A_{\omega_1} \cup B_\alpha \rangle$ and $w(\langle A_{\omega_1} \cup B_\alpha \rangle / A_{\omega_1}) = \aleph_0$, and

(4ii) $\langle A_{\omega_1} \cup B_\alpha \rangle$ is not a regular subalgebra of B_{ω_1}.

Condition $(4i)$ follows from $(0c)$ and $(4ii)$ is $(6c)$ for $\beta = \omega_1$.

As to the moreover-assertion, it follows from $(1c)$, $(2c)$, and $(3c)$ that $(A_\alpha)_{\alpha < \omega_1}$ is a standard filtration of A_{ω_1} (all these are isomorphic) such that $A_\alpha \leq_{free} B_{\omega_1}$ for all α. $\square$

For the application in section 4 we need twisted embeddings of higher cardinality. In the next proof we construct them simply by modifying the one we have got so far.

Proposition 6.2.5 (Fuchino, unpublished) *Consider a filtration $(C_\alpha)_{\alpha < \omega_1}$ of a free Boolean algebra C, where C_0 is infinite and free and $C_\alpha \leq_{free} C_{\alpha+1}$ for all α. Then there exists a twisted extension $C \leq C'$ such that $C_\alpha \leq_{free} C'$ for all α.*

Proof. By the definition 1.4.1 of $\leq_{free}$, the conditions imply that we can find pairwise disjoint infinite sets $X_{-1}, X_0, \ldots, X_\alpha \ldots$ of free generators of C such that $C_\alpha = \langle \bigcup_{\beta < \alpha} X_\beta \rangle$. ($\beta = -1$ is considered to make C_0 infinite.) We split $X_\alpha = Y_\alpha \cup Z_\alpha$, in such a way that each Y_α is countably infinite and $Z_\alpha \neq \emptyset$. Then we put

$$D_\alpha = \langle \bigcup_{\beta < \alpha} Y_\beta \rangle, \ E_\alpha = \langle \bigcup_{\beta < \alpha} Z_\beta \rangle, \ D = \langle \bigcup_{\alpha < \omega_1} Y_\alpha \rangle \ \text{and} \ E = \langle \bigcup_{\alpha < \omega_1} Z_\alpha \rangle.$$

Then (with a little abuse of notation)

$$C = D \otimes E = \bigcup_{\alpha < \omega_1} [D_\alpha \otimes E_\alpha].$$

Notice that $(D_\alpha)_{\alpha < \omega_1}$ is isomorphic to the standard filtration of $D \simeq \mathrm{Fr}\,\omega_1$. Therefore, by the previous proposition, there exists a twisted extension $D \leq B$ such that $D_\alpha \leq_{free} B$ for all α. We put $C' = B \otimes E$ and verify the conditions of twistedness.

(1) C' is free because $B \simeq \mathrm{Fr}\,\omega_1$ and E are both free.

(2) From $D \leq_{rc} B$ we immediately get $C = D \otimes E \leq_{rc} B \otimes E = C'$.

(3) Assume that $W \subseteq C'$ is countable and such that $C \cup W$ is dense in $C' \restriction c' = \bigvee_{i=1}^n b_i \otimes e_i$. Then the set W^* of all elements of B 'mentioned' in elements of

$W \subseteq B \otimes E$ is also countable. An easy verification shows that $D \cup W^*$ is dense in $B \upharpoonright b_1$, which contradicts the twistedness of $D \leq B$.

(4) Let $(B_\alpha)_{\alpha < \omega_1}$ be the twisted filtration of B over D. Then $(B_\alpha \otimes E)_{\alpha < \omega_1}$ is a filtration of C' and, for each α

(4i) $D \otimes E \leq B_\alpha \otimes E$ and $w(B_\alpha \otimes E / D \otimes E) = w(B_\alpha / D) = \aleph_0$, and

(4ii) $B_\alpha \otimes E$ is not regular in $B \otimes E$ (easy and left to the reader).

Moreover, from $D_\alpha \leq_{free} B$ and $E_\alpha \leq_{free} E$ we immediately get

$$C_\alpha = D_\alpha \otimes E_\alpha \leq_{free} B \otimes E = C'. \quad \square$$

The existence of twisted embeddings is now settled. What are they good for? As we mentioned above, they look very much like standard embeddings $\operatorname{Fr} \kappa \leq \operatorname{Fr} \kappa \otimes \operatorname{Fr} \omega_1$. But they are not even co-complete with standard embeddings. This is the special case $B = C$ of the following proposition, which will be used in section 3.

Proposition 6.2.6 *Consider Boolean algebras and embeddings as shown in the diagram*

$$
\begin{array}{ccccc}
A & \leq_{rc} & B & \leq_{rc} & C \\
\wedge|d & & & & \wedge|d \\
D & & \leq & & E \\
\vee|d & & & & \vee|d \\
F & & \leq_{rc} & & F'
\end{array}
$$

If F is free and $F \leq_{free} F'$, then $A \leq B$ cannot be twisted.

Proof.   Excluding trivial cases, we may assume that A and B are free and $w(B/A) = \aleph_1$. Take any subalgebra $X \leq B$ of power $\aleph_1$ such that $B = \langle A \cup X \rangle$. Let $(X_\alpha)_{\alpha < \omega_1}$ be a filtration of X consisting of countable subalgebras. We put $B_\alpha = \langle A \cup X_\alpha \rangle$. Then $(B_\alpha)_{\alpha < \omega_1}$ is a filtration of B over A such that $w(B_\alpha / A) = \aleph_0$ for all α. By 6.2.2, it is sufficient to show that

(*) $B_\alpha \leq_{reg} B$ for a closed and unbounded set of indices α.

For each $x \in X$ we let $H(x)$ denote a maximal subset of pairwise disjoint elements of $F' \upharpoonright x$. Because of the ccc, all $H(x)$ are countable and $\bigcup_{x \in X} H(x)$ has power $\aleph_1$. As $F \leq_{free} F'$ we can take a free subalgebra $G \leq F'$ of power $\aleph_1$ which is independent of F and such that $\bigcup_{x \in X} H(x) \subseteq \langle F \cup G \rangle \leq_{free} F'$.

CLAIM 1. *A and G are independent subalgebras of E.*

Indeed, for $0 < a \in A$ and $0 < g \in G$ we may take $0 < f \leq a$, $f \in F$. As G and F are independent, we get $a \wedge g \geq f \wedge g > 0$.

CLAIM 2. *Consider $a \in A$, $x \in X$, and $e \in F'$. If $a \wedge x \wedge e > 0$, then $f \wedge h \wedge e > 0$ for suitable $f \in F\!\restriction a$ and $h \in H(x)$.*

Indeed, first from $F' \leq_d E$ we get some $c \in F'$ such that $0 < c \leq a \wedge x \wedge e$. From $c \leq x$ we get, by maximality of $H(x)$, some $h \in H(x)$ such that $c \wedge h > 0$. Let $f \in F \leq_{rc} F'$ be minimal above $c \wedge h$. Then $0 < c \wedge h \leq f \wedge h \wedge e$.

It remains to see that $f \leq a$. If that were not the case, $0 < f - a \in D \geq_d F$, so $0 < f' \leq f - a$ for some $f' \in F$. Then $f - f' < f$, and $f' \wedge a = 0$ yields $f' \wedge c \wedge h = 0$, contradicting the minimality of f above $c \wedge h$.

Consider any element $g \in G \leq E \geq_d C \geq_{rc} B$. The ideal $B\!\restriction g$ is regular, which allows us to take a subset $K(g) \subseteq B$ such that $B\!\restriction g = K(g)^d$. Being free, B satisfies the ccc, and $K(g)$ can be taken countable.

Next we consider a standard filtration $G = \bigcup_{\alpha < \omega_1} G_\alpha$ of the free Boolean algebra G. Notice that $\langle F \cup G_\alpha \rangle \leq_{free} \langle F \cup G \rangle \leq_{free} F'$, hence $\langle F \cup G_\alpha \rangle \leq_{rc} F'$, for all α. It should be clear that the set of all $\alpha < \omega_1$ such that

(i) $H(x) \subseteq \langle F \cup G_\alpha \rangle$ for all $x \in X_\alpha$, and

(ii) $K(g) \subseteq B_\alpha = \langle A \cup X_\alpha \rangle$, for all $g \in G_\alpha$,

is closed and unbounded in ω_1.

CLAIM 3. *If α satisfies (i) and (ii), then for each $0 < e \in \langle F \cup G_\alpha \rangle$ there exists $0 < b \in B_\alpha$ such that $B_\alpha \!\restriction (b - e) = \{0\}$.*

To see this we write $e = \bigvee_{i=1}^n f_i \wedge g_i$, where $f_i \in F$, $g_i \in G_\alpha$, and all $f_i \wedge g_i > 0$.

Use density to take $a_1 \in A$ such that $0 < a_1 \leq f_1$. By claim 1, A and $G \supseteq G_\beta$ are independent, so $a_1 \wedge g_1 > 0$, i.e., $a_1 \not\in B\!\restriction - g_1 = K(-g_1)^d$, hence $a_1 \wedge k > 0$ for some $k \in K(-g_1) \subseteq B_\alpha$. The element $b = a_1 \wedge k \in B_\alpha$ is as required. Any $d \in B_\alpha\!\restriction (b - e)$ satisfies $d \leq b \leq a_1 \leq f_1$, $d \leq b \leq k$, and $0 = d \wedge e \geq d \wedge f_1 \wedge g_1 = d \wedge g_1$. So, $d \in B_\alpha\!\restriction - g_1 \subseteq B\!\restriction - g_1 = K(-g_1)^d$, hence $0 = d \wedge k = b$.

Combining claims 2 and 3, we finally show that for all α satisfying (i) and (ii) we must have $B_\alpha \leq_{reg} E$, hence $B_\alpha \leq_{reg} B$. This yields (*) and will thus end the proof of the proposition.

Let $e \in E \setminus \{0\}$ be given. Choose a non-zero $e_1 \in F'$ such that $e_1 \leq e$. Take $e_2 \in \langle F \cup G_\alpha \rangle \leq_{rc} F'$ minimal above e_1. Use claim 3 to choose $b \in B_\alpha$ such that

$$B_\alpha \!\restriction (b - e_2) = \{0\}.$$

We check that $B_\alpha \!\restriction (b - e) = \{0\}$.

Let $w \in B_\alpha = \langle A \cup X_\alpha \rangle$ be given such that $w \leq b$ and $w \wedge e = 0$. We have to prove that $w = 0$. Without loss we can assume that $w = a \wedge x$ for some $a \in A$, $x \in X_\alpha$. By the choice of b, it is sufficient to check that $0 = w \wedge e_2 = a \wedge x \wedge e_2$. If this were not true, we could use claim 2 to find $f \in F\!\restriction a$ and $h \in H(x) \subseteq G_\alpha$ such that $f \wedge h \wedge e_2 > 0$. On the other hand, $f \wedge h \leq a \wedge x = w$, so that $f \wedge h \wedge e_1 = 0$. Hence $f \wedge h \wedge e_2 = 0$, since e_2 is minimal above e_1 in $\langle F \cup G_\alpha \rangle \ni f \wedge h$. □

6.3 An rc-filtered Boolean algebra which is not weakly projective

To construct the example mentioned in the title, we iterate a twisted embedding ω_2 times.

More precisely, we let A be the union of a continuous chain $(A_\alpha)_{\alpha < \omega_2}$ of free Boolean algebras of power $\aleph_1$ such that all embeddings $A_\alpha \leq A_{\alpha+1}$ at successor stages are twisted. To see that such a chain exists, we must convince ourselves that at limit stages $A_\lambda = \bigcup_{\alpha < \lambda} A_\alpha$ remains free. Twisted embeddings being relatively complete and the chain continuous, A_λ is rc-filtered, by 2.2.8. As it has power $\aleph_1$ it is therefore projective and Ščepin's Theorem 1.4.11 applies, once we know that all ultrafilters have character $\aleph_1$. This follows immediately from the lemma below, applied to $A_0 \leq_{rc} A_\lambda$. Indeed, any countably generated ultrafilter of A_λ would give rise to a countably generated ultrafilter of $A_0 \simeq \mathrm{Fr}\,\omega_1$.

Lemma 6.3.1 *Assume $B \leq_{rc} C$ and let U be an ultrafilter of C. If $A \leq C$ is closed under the upper projection $q : C \to B$ and $A \cap U$ generates U, then $A \cap U \cap B$ generates $U \cap B$.*

Proof. We must find some $a \in A \cap U \cap B$ below any given $b \in U \cap B$. As $A \cap U$ generates U, there is $a' \in A \cap U$ below b. But then $a' \leq q(a') \leq b$ because $q(a')$ is minimal above a' in B. Hence $q(a') \in U$. As $q(a')$ is also in $A \cap B$, $a = q(a')$ has the required properties. □

We now know that the chain $(A_\alpha)_{\alpha < \omega_2}$ and its union A exist. By 2.2.8, A is rc-filtered. It is not hard to see that each ultrafilter of A has character $\aleph_2$. But Ščepin's theorem no longer applies, because there is no reason why A should be projective. The rest of this section will be devoted to proving

Proposition 6.3.2 *The rc-filtered algebra A is not co-complete with a projective Boolean algebra.*

Proof. Notice first that $\pi(A \restriction a) = \aleph_2$ for all non-zero $a \in A$. Indeed, if $V \subseteq A$ has power $\leq \aleph_1$ then $\{a\} \cup V \subseteq A_\alpha$ for some α. As $A_\alpha \leq A_{\alpha+1}$ is twisted, it is not even true that A_α is dense in $A_{\alpha+1} \restriction a$, let alone V in $A \restriction a$.

So if A were co-complete with some projective Boolean algebra, then it were co-complete with $\mathrm{Fr}\,\omega_2$, by 5.2.3. In other words, the completion of A would contain a free dense subalgebra of power $\aleph_2$. We derive a contradiction from assuming

$$A \leq_d E \geq_d F \simeq \mathrm{Fr}\,\omega_2$$

for some E.

Let $(F_\alpha)_{\alpha < \omega_2}$ be a filtration of F such that all F_α are free of power $\aleph_1$ and $F_\alpha \leq_{free} F_\beta$ for all $\alpha < \beta$. An easy argument shows that there is a closed and unbounded, hence non-empty, set of indices α such that

$$A_\alpha \leq_d \langle A_\alpha \cup F_\alpha \rangle \geq_d F_\alpha.$$

Fix such an ordinal and consider the diagram

$$A_\alpha \quad \leq_{rc} \quad A_{\alpha+1} \quad \leq_{rc} \quad A$$
$$\wedge|_d \qquad\qquad\qquad \wedge|_d$$
$$\langle A_\alpha \cup F_\alpha \rangle \qquad \leq \qquad E$$
$$\vee|^d \qquad\qquad\qquad \vee|^d$$
$$F_\alpha \qquad\qquad \leq_{rc} \qquad F$$

and notice that we are in the conditions of 6.2.6. It follows that $A_\alpha \leq A_{\alpha+1}$ is not twisted, contradiction. $\square$

6.4   On the number of rc-filtered Boolean algebras

It is convenient to introduce the notation $I(\textit{rc-filtered}, \kappa)$ for the maximal number of pairwise non-isomorphic rc-filtered Boolean algebras of power κ. The behaviour of the corresponding function $I(\textit{projective}, \kappa)$ is completely settled, thanks to results of Koppelberg's (cf. section 4 of [35]):

$$I(\textit{projective}, \kappa) = \begin{cases} 2^{\aleph_0} & , \quad \kappa = \aleph_0 \\ 2^{<\kappa} & , \quad \kappa \text{ regular uncountable} \\ 2^\kappa & , \quad \kappa \text{ singular uncountable} \end{cases}$$

From 2.2.7 and 2.2.6 it then follows that

$$I(\textit{rc-filtered}, \kappa) = \begin{cases} 2^{\aleph_0} & , \quad \kappa = \aleph_0 \\ 2^{\aleph_0} & , \quad \kappa = \aleph_1 \\ 2^\kappa & , \quad \kappa \text{ singular uncountable} \end{cases}$$

In this section we prove the following unpublished result of Fuchino's.

Theorem 6.4.1 *Assume that the regular cardinal number $\kappa \geq \aleph_2$ contains a stationary subset S such that*

(s1) $cf(\alpha) = \omega_1$, for all $\alpha \in S$, and

(s2) $\lambda \cap S$ is not stationary in λ, for all limit ordinals $\lambda < \kappa$.

Then $I(\textit{rc-filtered}, \kappa) = 2^\kappa$.

Below we call a stationary subset $S \subseteq \kappa$ *special* if it satisfies (s1) and (s2).

Corollary 6.4.2 (1) $I(\textit{rc-filtered}, \aleph_2) = 2^{\aleph_2}$
(2) *Assume $V=L$. Then $I(\textit{rc-filtered}, \kappa) = 2^\kappa$, for all $\kappa \geq \aleph_2$ that are not weakly compact.*

Proof of the corollary. (1) The natural choice

$$S = \{\alpha < \omega_2 : cf(\alpha) = \omega_1\}$$

turns out to be special. To verify (s2), we distinguish two cases. If $cf(\lambda) = \omega$ and $(\alpha_n)_{n<\omega}$ is closed and unbounded in λ, then $(\alpha_n + 1)_{n<\omega}$ is also closed and unbounded in λ and disjoint from S.

If $cf(\lambda) = \omega_1$ and $(\alpha_\beta)_{\beta<\omega_1}$ is closed and unbounded in λ, then $\{\alpha_\beta : cf(\beta) = \omega\}$ is also closed and unbounded and, as $cf(\alpha_\beta) = cf(\beta)$, none of its elements belongs to S.

(2) It is a well-known (and difficult) theorem of Jensen's (6.1 in [32] see also 11.1 in [8]), that, under V=L, for a regular cardinal to be weakly compact is equivalent to the non-existence of a special stationary set. We will not even define what weakly compact means but only mention that such cardinals are very big: there are strongly inaccessibly and even have inaccessibly many inaccessibles below them. $\square$

The **proof** of the theorem proceeds in several steps. First we manufacture many special sets out of the given one. Using Solovay's Theorem we may split S into κ pairwise disjoint stationary sets: $S = \bigcup_{\alpha<\kappa} S_\alpha$. Then we put $T_A = \bigcup_{\alpha \in A} S_\alpha$ for each $\emptyset \neq A \subseteq \kappa$. This gives us 2^κ stationary subsets of κ which, being subsets of the original S, are all special. Moreover, for $A \neq B$, either $T_A \setminus T_B$ or $T_B \setminus T_A$ is stationary.

In the second step we code a given special stationary T in an rc-filtered Boolean algebra C^T of power κ. To this end, a continuous rc-chain $(C_\alpha^T)_{\alpha<\kappa}$ will be constructed such that

(1) All C_α^T are free; $C_0^T \simeq \operatorname{Fr}\omega_1$.

(2) $w(C_{\alpha+1}^T/C_\alpha^T) = \aleph_1$ (hence $|C_\alpha^T| < \kappa$, for all α)

(3) If $\alpha < \beta < \kappa$ and $\alpha \notin T$, then $C_\alpha^T \leq_{free} C_\beta^T$.

(4) If $\alpha \in T$, then $C_\alpha^T \leq C_{\alpha+1}^T$ is a twisted embedding.

As long as T is fixed, we drop the superscript.
We START with $C_0 = \operatorname{Fr}\omega_1$. As continuity is required at LIMIT STEPS we have to put

$$C_\lambda = \bigcup_{\alpha<\lambda} C_\alpha = \bigcup_{\beta<cf(\lambda)} C_{\alpha_\beta},$$

where $(\alpha_\beta)_{\beta<cf(\lambda)}$ is closed and unbounded in λ and contains no elements of T.

C_λ is free since C_{α_0} is free and $C_{\alpha_\beta} \leq_{free} C_{\alpha_{\beta+1}}$ for all β. Conditions (2) and (4) are vacuous and (3) easily follows from transitivity: $\alpha < \alpha_\beta$, for some β, hence $C_\alpha \leq_{free} C_\beta \leq_{free} C_\lambda$.
SUCCESSOR STEP for $\alpha \notin T$. We put $C_{\alpha+1} = C_\alpha \otimes \operatorname{Fr}\omega_1$ and obviously preserve all conditions.

The real important step is the SUCCESSOR STEP for $\alpha \in T$. All the work has been done already in section 2. As T is special, α must be a limit ordinal of cofinality ω_1. As condition (3) holds up to α, we can apply proposition 6.2.5 to the filtration $(C_{\alpha_\beta})_{\beta<\omega_1}$. It yields some $C_{\alpha+1}$ such that $C_\alpha \leq C_{\alpha+1}$ is twisted but $C_{\alpha_\beta} \leq_{free} C_{\alpha+1}$ for all β.

Conditions (1), (2), and (4) are then obvious. (3) follows from transitivity again. If $T \ni \gamma < \alpha$, then $\gamma < \alpha_\beta$ for some β, hence $C_\gamma \leq_{free} C_{\alpha_\beta}$, by induction hypothesis, and $C_{\alpha_\beta} \leq_{free} C_{\alpha+1}$, by construction.

The chain $(C_\alpha^T)_{\alpha<\kappa}$ is constructed. Putting $C^T = \bigcup_{\alpha<\kappa} C_\alpha^T$, we get an rc-filtered Boolean algebra of power κ.

In the last step of the proof we show that C^T and C^U are not isomorphic, provided that $T \setminus U$ is stationary. Assume, on the contrary, that $f : C^T \to C^U$ is an isomorphism. As usual, one verifies that the set of all α such that $f(C_\alpha^T) = C_\alpha^U$ is closed and unbounded. Take $\alpha < \beta$ in that set, such that $\alpha \in T \setminus U$. Then the embeddings $C_\alpha^T \leq C_\beta^T$ and $C_\alpha^U \leq C_\beta^U$ are isomorphic. From $\alpha \notin U$, we get $C_\alpha^U \leq_{free} C_\beta^U$, by construction, hence $C_\alpha^T \leq_{free} C_\beta^T$, by isomorphism. By the lemma below, $C_\alpha^T \leq C_{\alpha+1}^T$ cannot be twisted. On the other hand, $\alpha \in T$ implies that the embedding was constructed to be twisted. This contradiction ends the proof. $\square$

The following lemma is of course a special case of 6.2.6. To spare the readers who are not interested in co-completeness the more difficult argument above, we give the following easy proof.

Lemma 6.4.3 *Assume $A \leq_{rc} B \leq_{rc} \langle A \cup G \rangle$, where G is free and independent of A. Then $A \leq B$ cannot be twisted.*

Proof. We exclude trivialities by assuming that B is free and $w(B/A) = \aleph_1$. Let $(B_\alpha)_{\alpha<\omega_1}$ be any filtration of B such that $A \leq B_\alpha$ and $w(B_\alpha/A) = \aleph_0$. By 6.2.2 it is sufficient to check that

(*) $B_\alpha \leq_{reg} B$ for a closed and unbounded set of indices.

Replacing it if necessary by a suitable subset, we may assume that G has power $\aleph_1$. Let $(G_\alpha)_{\alpha<\omega_1}$ be a standard filtration of G. Let $q : \langle A \cup G \rangle \to B$ denote the upper projection, witnessing relative completeness. It should be clear that a closed and unbounded set of indices α satisfies

(i) $B_\alpha \leq \langle A \cup G_\alpha \rangle$,   and   (ii) $q(G_\alpha) \subseteq B_\alpha$.

CLAIM   *If α satisfies (i) and (ii), then $B_\alpha \leq_{reg} \langle A \cup G_\alpha \rangle$.*

Indeed, let $0 < c = \bigvee_{i=1}^n a_i \wedge g_i \in \langle A \cup G_\alpha \rangle$ be given, where $a_i \in A, g_i \in G_\alpha$, and, $a_i \wedge g_i > 0$. As $g_1 \leq q(g_1) \in B_\alpha$, we have $0 < a_1 \wedge g_1 \leq a_1 \wedge q(g_1) \in B_\alpha$. Moreover,

$$B_\alpha \restriction (a_1 \wedge q(g_1) - \bigvee_{i=1}^n a_i \wedge g_i) \subseteq B_\alpha \restriction (q(g_1) - g_1) \subseteq B \restriction (q(g_1) - g_1) = \{0\},$$

by definition of q. The claim is proved. By transitivity, $B_\alpha \leq_{reg} \langle A \cup G_\alpha \rangle \leq_{rc} \langle A \cup G \rangle \geq B$ implies $B_\alpha \leq_{reg} B$, as desired. $\square$

Note added in proof (September 1994)

Very recently S. Fuchino and S. Shelah have proved in ZFC that the number $I(rc\text{-}filtered, \kappa)$ equals 2^κ for all cardinal numbers $\kappa \geq \aleph_2$. This result will be contained in [21], which is currently being prepared by Fuchino.

Appendix A

Set-theoretic aspects of nearly projective Boolean algebras

by **Sakaé Fuchino**[1]

A.1   Elementary substructures of $\mathcal{H}_\chi$

In this section, we shall introduce a new method (that of elementary substructures) which will be used extensively in later sections. Some notions and facts from model theory and set theory we need for the method are also reviewed here. We shall try to make the following as self-contained as possible. For some notions from basic model theory which remain unexplained here, the reader may consult [7]. Some more details of set theoretic arguments can be found in [39], [30] and/or [31]. A good survey on applications of the method of elementary substructures to set-theoretic topology can be found in [9].

For the readers who are not familiar with set theory, we would like to emphasize that, from our set-theoretic stand-point, the whole of mathematics is thought to be done in the framework of axiomatic set theory. This means in particular that all mathematical objects are thought here to be (coded into) sets. For example, a mapping f from a set a to another set b is thought to be just a subset of $a \times b$ (intuitively the graph of f: $\{(u,v) : u \in a, v = f(u)\}$)

---

[1] This appendix was written while I was working at the Institute of Mathematics of Hebrew University Jerusalem. I would like to thank James Cummings for improving some of my awkward expressions in Japanese/German English in the text.

which satisfies the conditions

$$(\forall u \in a)(\exists v \in b)((u, v) \in f),$$

$$(\forall u \in a)(\forall v \in b)(\forall w \in b)((u, v) \in f \land (u, w) \in f \rightarrow v = w).$$

For more, see the proof of Lemma A.1.9 below.

The method of elementary substructures we are going to develop here has some similarity to non-standard analysis[2]: in non-standard analysis, we extend the 'universe' of mathematics (set theory) by adding non-standard objects, for example infinitesimal (imaginary) real numbers. We then carry on our discussion using these non-standard objects, and return to the original standard universe with conclusions obtained. In contrast to that, in our method of elementary substructures, we start from the universe of set theory (let us call it V), and find a 'small' substructure M of V which is quite 'similar' to V. Seen from inside M, the objects in V which are not in M are like non-standard objects. Hence we can study how M behaves also using these non-standard objects outside M and establish that some statement holds in M. Then, since M was similar to V, we conclude that this statement holds also in V, that is, it is true[3].

Unfortunately there is one fatal problem in the method described as above. Namely, the whole set-theoretic universe is so to speak 'too big' to have such M quite similar to it. Actually from the existence of such M we would be able to conclude the consistency of set theory. But by Gödel's Incompleteness Theorem we can not prove the consistency of set theory inside set theory itself! To overcome this difficulty, we start not from the whole universe V of set theory but from a large enough substructure ($\mathcal{H}_\chi$ as defined below for a large enough cardinal χ) of V which has enough similarity to V.

To define $\mathcal{H}_\chi$, we need the notion of transitive sets: x (a set or a proper class) is said to be *transitive*, if for every $y \in x$ and $z \in y$, $z \in x$ holds. For any set x, let $\mathrm{trcl}(x)$ (the *transitive closure of X*) be the smallest set u including x such that u is transitive. Formally $\mathrm{trcl}(x)$ can be defined as follows: for $n \in \omega$ let $tc_n(x)$ be defined inductively by $tc_0(x) = x$ and $tc_{n+1}(x) = tc_n(x) \cup \{ z : z \in y$ for some $y \in tc_n(x) \}$. Then, let $\mathrm{trcl}(x) = \bigcup_{n \in \omega} tc_n(x)$.

Lemma A.1.1 *If $x \in y$ then $\mathrm{trcl}(x) \subseteq \mathrm{trcl}(y)$.*

Proof If $x \in y$ then $tc_n(x) \in tc_{n+1}(y)$ holds for every $n \in \omega$. Since $(tc_n(y))_{n \in \omega}$ is an increasing sequence, it follows that that $\mathrm{trcl}(x) = \bigcup_{n \in \omega} tc_n(x) \subseteq \bigcup_{n \in \omega} tc_n(y) = \mathrm{trcl}(y)$. $\qquad \square$ (Lemma A.1.1)

For a cardinal χ, let

$$\mathcal{H}_\chi = \{ x : |\mathrm{trcl}(x)| < \chi \}.$$

---

[2] The following explanation in this paragraph is just a very intuitive one (*i.e.* almost a cheat).

[3] A good example of this type of argument is the proof of the Δ-System Lemma (Theorem A.1.13) below. Our application of the method of elementary substructure in later sections is actually a bit different from the description above — we also use the facts that there are 'many' (closed unbounded many) M as above, and that for an algebraic structure B in M, $B \cap M$ is a subalgebra of B.

Elements of $\mathcal{H}_\chi$ are said to be hereditarily of cardinality $< \chi$.

Lemma A.1.2 *For every cardinal χ, $\mathcal{H}_\chi$ is transitive.*

Proof If $x \in \mathcal{H}$ and $y \in x$ then $\mathrm{trcl}(y) \subseteq \mathrm{trcl}(x)$ by Lemma A.1.1. It follows that $|\mathrm{trcl}(y)| \leq |\mathrm{trcl}(x)| < \chi$. Hence we have $y \in \mathcal{H}_\chi$. ☐ (Lemma A.1.2)

$\mathcal{H}_\chi$ is actually a set (*i.e.* not a proper class). This can be seen in the following:

Lemma A.1.3 *For every cardinal χ, $\mathcal{H}_\chi \subseteq V_\chi$ holds where $(V_\alpha)_{\alpha \in \mathrm{Ord}}$ is the von Neumann's hierarchy[4].*

$(V_\alpha)_{\alpha \in \mathrm{Ord}}$ is defined inductively by:

$$V_0 = \emptyset;$$
$$V_{\alpha+1} = V_\alpha \cup P(V_\alpha);$$
$$V_\delta = \bigcup_{\alpha < \delta} V_\alpha \text{ for a limit ordinal } \delta.$$

By the inductive definition we see immediately that each V_α is a set. Hence from Lemma A.1.3 we can conclude that, as a subclass of a set V_χ, $\mathcal{H}_\chi$ is also a set.

Proof of Lemma A.1.3 It is enough to show that the claim holds for every regular χ. Suppose on the contrary that χ is regular and $\mathcal{H}_\chi \not\subseteq V_\chi$. Let $x \in \mathcal{H}_\chi \setminus V_\chi$ be of minimal $\in$-rank. Since every $y \in x$ is in $\mathcal{H}_\chi$ by Lemma A.1.2, and since $y \in x$ has smaller $\in$-rank than that of x, we have $x \subseteq V_\chi$. Since χ is regular and $|x| \leq |\mathrm{trcl}(x)| < \chi$, there is some $\beta < \chi$ such that $x \subseteq V_\beta$. But then $x \in V_{\beta+1} \subseteq V_\chi$. This is a contradiction. ☐ (Lemma A.1.3)

The following Theorem A.1.4 guarantees that there are many 'small' M which are quite similar to $\mathcal{H}_\chi$. Let us first define what we mean by the similarity of M to $\mathcal{H}_\chi$ in our context.

We are mainly interested here in structures of the form $M = (M, E)$ where E is a binary relation, though Theorems A.1.4, A.1.5, A.1.6 and so on also hold for arbitrary structures with countably many constants, functions and relations. Hence when we say simply "*structure*" we shall mean a structure of the form (M, E). If we are talking about structures in the more general sense with $\leq \aleph_0$ many constants, functions and relations we shall refer to them as "*algebraic structures*". We denote the formal (first order) language corresponding to structures of the form $M = (M, E)$ by 'L'. We shall write the binary relation symbol corresponding to the relation E also 'E'. For simplicity, we shall assume that formally every L-formula is built up from formulas of the form '$x \, E \, y$' or '$x = y$'(atomic formulas) by logical connectives '$\wedge$', '$\neg$' and quantifier '$\exists$'. As is well known, this is enough since '$\vee$', '$\rightarrow$' and '$\forall$' can be expressed by appropriate combinations of '$\wedge$', '$\neg$' and '$\exists$'. In our context, the binary relation E on M is often defined by $\{(a, b) \in M^2 \; : \; a \in b\}$. If this is the case, we shall write $M = (M, \in)$ and denote the corresponding binary relation symbol also by '$\in$'.

---

[4] Ord denotes the class of ordinals.

For a structure M, an L-formula $\varphi(x_1, \ldots, x_n)$ and elements $a_1, \ldots, a_n$ of M the relation "$M \models \varphi[a_1, \ldots, a_n]$" is intuitively the one which says that "φ holds for $a_1, \ldots, a_n$ inside M". Formally the relation is introduced by the inductive definition:

If φ is '$x \in y$' then
$$M \models \varphi[a_1, a_2] \Leftrightarrow a_1 \in a_2;$$

If φ is '$x = y$' then
$$M \models \varphi[a_1, a_2] \Leftrightarrow a_1 = a_2;$$

If φ is '$\psi_1 \wedge \psi_2$' then
$$M \models \varphi[a_1, \ldots, a_n] \Leftrightarrow M \models \psi_1[a_1, \ldots, a_n] \text{ and } M \models \psi_2[a_1, \ldots, a_n];$$

If φ is '$\neg\psi$' then
$$M \models \varphi[a_1, \ldots, a_n] \Leftrightarrow M \models \psi[a_1, \ldots, a_2] \text{ does not hold;}$$

If φ is '$\exists x \psi$' then
$$M \models \varphi[a_1, \ldots, a_n] \Leftrightarrow \text{ there is } a \in M \text{ such that}$$
$$M \models \psi[a, a_1, \ldots, a_n] \text{ holds.}$$

A substructure N of M is called an *elementary substructure of M* (notation: $N \prec M$) if for every L-formula $\varphi(x_1, \ldots, x_n)$ with n free variables and for every $a_1, \ldots, a_n \in N$
$$M \models \varphi[a_1, \ldots, a_n] \Leftrightarrow N \models \varphi[a_1, \ldots, a_n]$$

holds. It is easy to see that "$\prec$" is a transitive relation on structures. Also it is easily seen from the definition of '$\prec$' that if $M_0, M_1 \prec M$ and $M_0 \subseteq M_1$ then we have $M_0 \prec M_1$.

Theorem A.1.4 *a)* (Downward Löwenheim-Skolem Theorem) *For any $X \subseteq M$, there exists $N \prec M$ such that $X \subseteq N$ and $|X| = |N|$.*[5]

b) (Union of Chains) *If $(N_\alpha)_{\alpha < \delta}$ is an increasing sequence of elementary substructures of a structure M then $\bigcup_{\alpha < \delta} N_\alpha \prec M$.*

Proof *a)*: This will be proved after Theorem A.1.6.

b): Let $N = \bigcup_{\alpha < \delta} N_\alpha$. We show

$(*)_\varphi$ *for every $a_1, \ldots, a_n \in N$,*

$$M \models \varphi[a_1, \ldots, a_n] \Leftrightarrow N \models \varphi[a_1, \ldots, a_n]$$

by induction on φ. If φ is atomic, this is trivial. Also if φ is one of $\psi_1 \wedge \psi_2$, $\neg\psi_1$ and we have already shown $(*)_{\psi_1}$ and $(*)_{\psi_2}$, it is easy to show that $(*)_\varphi$ also holds. Now assume that φ is $\exists x \psi$ for an L-formula $\psi = \psi(x, x_1, \ldots, x_n)$ and we have already shown that $(*)_\psi$. Let $a_1, \ldots, a_n \in N$. First assume that $M \models \exists x \psi[x, a_1, \ldots, a_n]$. Let $\gamma < \delta$ be such that $a_1, \ldots, a_n \in N_\gamma$. Since $N_\gamma \prec M$, it follows that $N_\gamma \models \exists x \psi[x, a_1, \ldots, a_n]$ holds. Hence there is $a_0 \in N_\gamma$ such that $N_\gamma \models \psi[a_0, a_1, \ldots, a_n]$ holds. Again by $N_\gamma \prec M$, it follows that

---

[5] Recall that $|X|$ is defined as the maximum of $\aleph_0$ and the cardinality of X.

$M \models \psi[a_0, a_1, \ldots, a_n]$ holds. Hence by the induction hypothesis $(*)_\psi$, we have $N \models \psi[a_0, a_1, \ldots, a_n]$. It follows that $N \models \exists x \psi[x, a_1, \ldots, a_n]$. Now assume that $N \models \exists x \psi[x, a_1, \ldots, a_n]$. Then there is some $a_0 \in N$ such that $N \models \psi[a_0, a_1, \ldots, a_n]$. By $(*)_\psi$ it follows that $M \models \psi[a_0, a_1, \ldots, a_n]$. Hence we have that $M \models \exists x \psi[x, a_1, \ldots, a_n]$. $\qquad \Box$ (Theorem A.1.4)

By Theorem A.1.4, we can easily obtain the following generalization of Iwamura's Lemma (Lemma 2.1.6):

Theorem A.1.5 *Every uncountable structure M is the union of a continuous (well-ordered) chain of elementary substructures of M of cardinality strictly less than* $|M|$. $\qquad \Box$

The elementary substructure N of M in Theorem A.1.4,a) can be actually taken "uniformly" in the following sense. We say f is a *function on* M if $f : M^n \rightarrow M$ for some $n \in \omega$ (For $n = 0$, the 0-place function f is considered to be just an element of M).

Theorem A.1.6 (Skolem Functions) *For any structure M there exists a countable set F of functions on M such that for every substructure $N \subseteq M$, if N is closed with respect to every $f \in F$ (i.e. every 0-place $f \in F$ we have $f \in N$ and for every n-place $f \in F$ for $n \geq 1$ and $a_1, \ldots, a_n \in N$ we have $f(a_1, \ldots, a_n) \in N$) then $N \prec M$.*

Proof Let $\sqsubseteq$ be any well-ordering on M and let m_0 be the smallest element of M with respect to $\sqsubseteq$. For each L-formula $\varphi(x_0, x_1, \ldots, x_n)$, let $f_\varphi : M^n \rightarrow M$ be the function on M defined by

$$f_\varphi(a_1, \ldots, a_n) = \begin{cases} \text{the smallest } a_0 \in M \text{ (with respect to } \sqsubseteq \text{) such that} \\ M \models \varphi[a_0, a_1, \ldots, a_n] \text{ if such } a_0 \text{ exists;} \\[2mm] m_0; \text{ otherwise.} \end{cases}$$

We show that $F = \{f_\varphi : \varphi \text{ is an } L\text{-formula}\}$ is as desired. Let $N \subseteq M$ be closed with respect to f_φ for every L-formula φ. As in the proof of Theorem A.1.4,b), we show that $(*)_\varphi$ holds for every L-formula φ by induction on φ. If φ is atomic, this is clear. Also if φ is one of $\psi_1 \wedge \psi_2$, $\neg \psi_1$ and $(*)_{\psi_1}$, and $(*)_{\psi_2}$ hold, it is easy to show that $(*)_\varphi$ also holds. So assume that φ is $\exists x \psi$ and that we have already shown that $(*)_\psi$ holds. Let $a_1, \ldots, a_n \in N$. If $M \models \exists x \psi[x, a_1, \ldots, a_n]$, then by the definition of f_ψ, we have $M \models \psi[f_\psi(a_1, \ldots, a_n), a_1, \ldots, a_n]$. By $(*)_\psi$ and $f_\psi(a_1, \ldots, a_n) \in N$, it follows that $N \models \psi[f_\psi(a_1, \ldots, a_n), a_1, \ldots, a_n]$. Hence $N \models \exists x \psi[x, a_1, \ldots, a_n]$. Conversely if $N \models \exists x \psi[x, a_1, \ldots, a_n]$, then just as in the proof of Theorem A.1.4,b), we can show that $M \models \exists x \psi[x, a_1, \ldots, a_n]$. $\qquad \Box$ (Theorem A.1.6)

For a structure M let us call a countable set F as in Theorem A.1.6, a set of Skolem functions. For a (countable) set of Skolem functions F on M and any $X \subseteq M$, let us denote by $\bar{h}_F(X)$ the *Skolem hull of* X *with respect to* F, that is, the smallest set containing X closed with respect to every $f \in F$. For every $X \subseteq M$, we have $|X| = \left| \tilde{h}_F(X) \right|$ and $\tilde{h}_F(X) \prec M$ holds by Theorem A.1.6. Thus we obtained a proof of Theorem A.1.4, a).

For a set X and a cardinal κ, let $[X]^\kappa = \{ Y \subseteq X : |Y| = \kappa \}$. Similarly let $[X]^{<\kappa} = \{ Y \subseteq X : |Y| < \kappa \}$. $\mathcal{C} \subseteq [X]^\kappa$ is said to be *closed unbounded* if $\mathcal{C}$ is cofinal in $([X]^\kappa, \subseteq)$ and closed with respect to union of increasing sequences (with respect to $\subseteq$) of length $\leq \kappa$. Likewise $\mathcal{C} \subseteq [X]^{<\kappa}$ is said to be *closed unbounded* if $\mathcal{C}$ is cofinal in $([X]^{<\kappa}, \subseteq)$ and closed with respect to union of increasing sequences of length $< \kappa$.

Theorem A.1.7 *Let $X \subseteq M$, $x_0, \ldots, x_{n-1} \in M$ and let κ be a regular cardinal $\leq |X|^+$, then*

$$S = \{ X \cap N : N \prec M, \, x_0, \ldots, x_{n-1} \in M, \, |N| < \kappa \}$$

contains a closed and unbounded subset of $[X]^{<\kappa}$.

Proof Let F be a set of Skolem functions on M. Without loss of generality, we may assume that $x_0, \ldots, x_{n-1} \in \tilde{h}_F(\emptyset)$ (otherwise replace F with $F \cup \{ x_0, \ldots, x_{n-1} \}$). Let

$$C = \{ Y \subseteq X : |Y| < \kappa, \, X \cap \tilde{h}_F(Y) = Y \}.$$

Clearly $\mathcal{C} \subseteq S$. We show that $\mathcal{C}$ is closed unbounded in $[X]^{<\kappa}$: for any $Z \in [X]^{<\kappa}$ let $Y = X \cap \tilde{h}_F(Z)$. Then $|Y| < \kappa$ and since $Z \subseteq Y \subseteq \tilde{h}_F(Z)$, we have $\tilde{h}_F(Y) = \tilde{h}_F(Z)$. It follows that $X \cap \tilde{h}_F(Y) = X \cap \tilde{h}_F(Z) = Y$. Hence $Y \in \mathcal{C}$. If $(Y_\alpha)_{\alpha < \lambda}$ is an increasing sequence in $\mathcal{C}$ for some $\lambda < \kappa$ then $Y = \bigcup_{\alpha < \lambda} Y_\alpha$ is still of cardinality $< \kappa$. Since $\bigcup_{\alpha < \lambda} \tilde{h}_F(Y_\alpha)$ is closed under every $f \in F$, we have $\tilde{h}_F(Y) = \bigcup_{\alpha < \lambda} \tilde{h}_F(Y_\alpha)$. It follows that

$$X \cap \tilde{h}_F(Y) = X \cap (\bigcup_{\alpha < \lambda} \tilde{h}_F(Y_\alpha)) = \bigcup_{\alpha < \lambda} (X \cap \tilde{h}_F(Y_\alpha)) = \bigcup_{\alpha < \lambda} Y_\alpha = Y.$$

Hence $Y \in \mathcal{C}$. $\square$ (Theorem A.1.7)

Elementary substructures of $(\mathcal{H}_\chi, \in)$ of small cardinality are the structures we actually meant as we were talking about 'M' in the intuitive explanation we gave at the beginning of this section. From now on, we shall regard $\mathcal{H}_\chi$ always as the structure $(\mathcal{H}_\chi, \in)$. $\mathcal{H}_\chi$ (with the $\in$-relation) is almost a model of set theory: if χ is regular uncountable, $\mathcal{H}_\chi$ satisfies every axiom of set theory (ZFC) possibly except the power set axiom. $\mathcal{H}_\chi$ satisfies also the power set axiom if and only if χ is inaccessible (see [39]). However these facts will not be used in the following. What we need here is rather the fact that $\mathcal{H}_\chi$ is 'similar enough' to V in the sense that $\mathcal{H}_\chi$ is an 'elementary substructure' of V with respect to some properties (which are of course expressed by some L-formulas):

For a structure $M = (M, \in)$, an L-formula $\varphi(x_1, \ldots, x_n)$ is said to be *absolute over M* if for every $a_1, \ldots, a_n \in M$ we have:

$$\varphi(a_1, \ldots, a_n) \text{ holds } \Leftrightarrow M \models \varphi[a_1, \ldots, a_n].$$

In general it is not easy to decide whether a formula is absolute over some M and this can also depend on M. For example, if we have $\neg CH$ and φ is a formula expressing CH then φ is absolute over M if and only if CH does not hold in M. However, for our purpose, the restricted formulas defined below would suffice whose absoluteness over a transitive M will be shown in Lemma A.1.8.

For an L-formula $\varphi(x_0, x_1, \ldots, x_n)$, we denote by $(\exists x_0 \in x_1)\varphi(x_0, x_1, \ldots, x_n)$ the formula

$$\exists x_0(x_0 \in x_1 \wedge \varphi(x_0, x_1, \ldots, x_n)).$$

Similarly $(\forall x_0 \in x_1)\varphi(x_0, x_1, \ldots, x_n)$ is defined by

$$\forall x_0(x_0 \notin x_1 \vee \varphi(x_0, x_1, \ldots, x_n)).$$

Quantification of the form "$(\exists x_0 \in x_1)$" or "$(\forall x_0 \in x_1)$" is called *bounded quantification*. Note that $(\forall x_0 \in x_1)\varphi(x_0, x_1, \ldots, x_n)$ is equivalent to $\neg(\exists x_0 \in x_1)\neg\varphi(x_0, x_1, \ldots, x_n)$. An L-formula φ is said to be *restricted* if φ is built up from atomic formulas (formulas of the form '$x \in y$' or '$x = y$') by $\wedge$, $\neg$ and bounded quantification.

Lemma A.1.8 *If L-formula φ is restricted then for every transitive M, φ is absolute over M.*[6]

Proof We show that

$(^*_*)_\varphi$ *for every transitive M and $a_1, \ldots, a_n \in M$,*

$$\varphi(a_1, \ldots, a_n) \text{ holds} \iff M \models \varphi[a_1, \ldots, a_n]$$

holds for every restricted φ by induction on φ. If φ is atomic, this is clear. Also if φ is one of $\varphi_1 \wedge \varphi_1$, $\neg\varphi_1$ and $(^*_*)_{\varphi_1}$, and $(^*_*)_{\varphi_2}$ hold, it is easy to show that $(^*_*)_\varphi$ also holds. So assume that φ is $(\exists x \in x_1)\psi$ and we have already shown that $(^*_*)_\psi$ holds. Let M be an arbitrary transitive set and let $a_1, \ldots, a_n \in M$. If $(\exists x \in a_1)\psi(a_1, \ldots, a_n)$ then there is $a_0 \in a_1$ such that $\psi(a_0, a_1, \ldots, a_n)$. Since M is transitive we have $a_0 \in M$. By $(^*_*)_\psi$ it follows that $M \models \psi[a_0, a_1, \ldots, a_n]$. Hence we have $M \models (\exists x \in a_1)\psi[x, a_1, \ldots, a_n]$. Conversely, suppose that $M \models (\exists x \in a_1)\psi[x, a_1, \ldots, a_n]$. Then there is $a_0 \in M$ such that $M \models$ "$a_0 \in a_1 \wedge \psi[a_0, a_1, \ldots, a_n]$". By $(^*_*)_\psi$ it follows that $\psi(a_1, \ldots, a_n)$. Hence we have: $(\exists x \in a_1)\psi(x, a_1, \ldots, a_n)$. $\quad\square$ (Lemma A.1.8)

Lemma A.1.9 *The following properties can be expressed by restricted formulas and hence, by Lemma A.1.2 and Lemma A.1.8, they are absolute over $\mathcal{H}_\chi$ for any regular uncountable χ:* "$x = \{y, z\}$", "$x = (y, z)$", "$(x, y) \in z$" "$x \subseteq y \times z$", "$f : x \to y$", "$x = \text{dom}(f)$", "$x = \text{rng}(f)$", "$f(x) = y$", "f is a surjection from x to y", "f is an injection from x to y", "x is transitive", "x is an ordinal", "x is a limit ordinal" etc.*

---

[6] Note that φ and ψ in Lemma A.1.8 are "real" formulas outside set-theory while, in the definition and in the proof of Theorems A.1.4, A.1.6 etc. on elementary substructures, they are objects inside set-theory. This is because "φ holds" cannot be expressed by a set theoretic predicate with φ as a parameter unless set-theory is inconsistent.

Though the proof is rather lengthy, we shall demonstrate a large part of it in detail so that the reader may have the opportunity to see how some of mathematical notions are represented in the formal system of set theory.

Proof of Lemma A.1.9   "$x = \{y, z\}$" can be represented by the L-formula:

$$(\forall u \in x)(u = y \vee u = z) \wedge (\exists u \in x)(u = y) \wedge (\exists u \in x)(u = z).$$

For "$x = (y, z)$", recall that the ordered pair (a, b) is defined in set theory by $\{\{a\}, \{a, b\}\}$. Hence "$x = (y, z)$" can be represented by the L-formula:

$$(\forall u \in x)(\text{"}u = \{y, y\}\text{"} \vee \text{"}u = \{y, z\}\text{"}) \wedge$$
$$(\exists u \in x)(\text{"}u = \{y, y\}\text{"}) \wedge (\exists u \in x)(\text{"}u = \{y, z\}\text{"}).$$

"$(x, y) \in z$" can be represented by the L-formula:

$$(\exists u \in z)(\text{"}u = (x, y)\text{"}).$$

"$x \subseteq y \times z$" can be represented by the L-formula:

$$(\forall u \in x)(\exists v \in y)(\exists w \in z)(\text{"}u = (v, w)\text{"}).$$

As already seen on page 165, "$f : x \to y$" is represented by the L-formula:

$$\text{"}f \subseteq x \times y\text{"} \wedge (\forall u \in x)(\exists v \in y)(\text{"}(u, v) \in f\text{"}) \wedge$$
$$(\forall u \in x)(\forall v \in y)(\forall w \in y)(\text{"}(u, v) \in f\text{"} \wedge \text{"}(u, w) \in f\text{"} \to v = w).$$

"$x = \mathrm{dom}(f)$" is represented by the L-formula:

$$(\forall u \in x)(\exists v \in f)(\exists y \in v)(\exists z \in y)(\text{"}v = (u, z)\text{"}) \wedge$$
$$(\forall u \in f)(\exists v \in x)(\exists y \in u)(\exists z \in y)(\text{"}u = (v, z)\text{"}).$$

"$f(x) = y$" is represented by the L-formula:

$$(\exists u \in f)(\text{"}u = (x, y)\text{"}).$$

"x is transitive" is represented by the L-formula:

$$(\forall u \in x)(\forall v \in u)(v \in x).$$

"x is an ordinal" is represented by the L-formula:

$$\text{"}x \text{ is transitive"} \wedge (\forall u \in x)(\forall v \in x)(u \in v \vee v \in u \vee u = v).$$

"x is a limit ordinal" is represented by the L-formula:

$$\text{"}x \text{ is an ordinal"} \wedge (\forall u \in x)(\exists v \in x)(u \in v).$$

The rest of the proof is left to the reader.   ◻ (Lemma A.1.9)

To guarantee that the absoluteness of the properties like the ones listed in Lemma

A.1.9 above does not depend on the choice of formal definition of them, we should assume that M satisfies a portion of set theory, so that the equivalence between different possible formal definitions of each of these notions holds also in M. For $\mathcal{H}_\chi$ for regular χ this is actually the case. But instead of doing so, we simply assume here that all properties needed in the arguments below are introduced by possibly absolute formal definitions.

An example of not (necessarily) absolute property is "x is a cardinal". This property can be represented by the formula:

"x is an ordinal" $\wedge$ $(\forall y \in x)\neg((\exists f)($"$f$ is a surjection from y onto x"$))$.

Now for a transitive M if $\kappa \in M$ is a cardinal then $M \models$ "κ is a cardinal". On the other hand if $\alpha \in M$ is an ordinal which is not a cardinal then there is some $\beta \in \alpha$ and a surjection $f : \beta \to \alpha$. Though by transitivity we have $\beta \in M$ there is no reason that we also have $f \in M$. However in case of $M = \mathcal{H}_\chi$ we can show the absoluteness of the property over M: if $\alpha \in \mathcal{H}_\chi$ and we have a surjection $f : \beta \to \alpha$ for some $\beta \in \alpha$, then $|\mathrm{trcl}(f)| = |\alpha| < \chi$. Hence we have $f \in \mathcal{H}_\chi$.

The following argument will be used quite often:

Lemma A.1.10 *Let* $M \prec N$ *and* $a_1, \ldots, a_n \in M$. *If* $N \models \exists! x \varphi[x, a_1, \ldots, a_n]$ *holds*[7] *and* $a_0 \in N$ *satisfies* $N \models \varphi[a_0, a_1, \ldots, a_n]$, *then we have* $a_0 \in M$.

Proof By $M \prec N$, we have $M \models \exists x \varphi[x, a_1, \ldots, a_n]$. Let $b \in N$ be such that $N \models \varphi[b, a_1, \ldots, a_n]$. Then again by $M \prec N$ we have $M \models \varphi[b, a_1, \ldots, a_n]$. Hence by $M \models \exists! x \varphi[x, a_1, \ldots, a_n]$ it follows that $a_0 = b$. $\square$ (Lemma A.1.10)

To demonstrate how the absoluteness of some formulas is used in a proof by means of elementary substructures of $\mathcal{H}_\chi$, we shall prove the following lemma with some details which will be simply omitted in later proofs.

Lemma A.1.11 *Suppose* χ *is regular uncountable and* $M \prec H_\chi$.

a) $\omega \in M$ *and* $\omega \subseteq M$;

b) *If* $a \in M$ *and* $|a| \leq \aleph_0$ *then we have* $a \subseteq M$.

Proof a): "x is ω" is represented by the L-formula $\varphi(x)$ of the form:

"x is a limit ordinal" $\wedge$ $(\forall y \in x)($"y is not a limit ordinal"$)$.

Since $\mathrm{trcl}(\omega) = \omega$ is countable, we have $\omega \in \mathcal{H}_\chi$. By the absoluteness of $\varphi(x)$ over $\mathcal{H}_\chi$ and Lemma A.1.8, we have $\mathcal{H}_\chi \models \exists x \varphi(x)$. We can also show that $\mathcal{H}_\chi \models (\exists! x)\varphi(x)$. Hence by Lemma A.1.10, it follows that $\omega \in M$. By a similar argument we can show $0 \in M$ and $n \in M \Rightarrow n + 1 \in M$. This is because, in set theory, 0 is defined as empty set and $n + 1$ is the set $n \cup \{n\}$. Hence

---

[7] "$\exists! x \varphi(x, x_1, \ldots, x_n)$" is a formula saying that "there exists exactly one 'x' such that $\varphi(x, x_1, \ldots, x_n)$". Formally it is defined by:

$$\exists x \varphi(x, x_1, \ldots, x_n) \wedge \forall x \forall y (\varphi(x, x_1, \ldots, x_n) \wedge \varphi(y, x_1, \ldots, x_n) \to x = y).$$

Clearly we have: $\exists! x \varphi(x, x_1, \ldots, x_n) \to \exists x \varphi(x, x_1, \ldots, x_n)$.

it is easily shown that assertions "x is 0" and "$y = x + 1$" are represented by some restricted formulas $\psi_1(x)$, $\psi_2(x, y)$ and we have $\mathcal{H}_\chi \models (\exists!x)\psi_1(x)$ and $\mathcal{H}_\chi \models (\forall x)(\exists!y)\psi_2(x, y)$.

b): Since $a \in \mathcal{H}_\chi$ and $|a| \leq \aleph_0$, there is a surjection $f_0 : \omega \to a$. Clearly $f_0 \in \mathcal{H}_\chi$. Hence we have

$$\mathcal{H}_\chi \models \text{"}\exists f(f \text{ is a surjection from } \omega \text{ onto } a)\text{"}.$$

By $M \prec \mathcal{H}_\chi$ it follows that

$$M \models \text{"}\exists f(f \text{ is a surjection from } \omega \text{ onto } a)\text{"}.$$

Hence there is $f_1 \in M$ such that

$$M \models \text{"} f_1 \text{ is a surjection from } \omega \text{ onto } a\text{"}.$$

Again by $M \prec \mathcal{H}_\chi$, it follows that

$$\mathcal{H}_\chi \models \text{"} f_1 \text{ is a surjection from } \omega \text{ onto } a\text{"}.$$

By the absoluteness of the assertion "f is a surjection from x onto y" over $\mathcal{H}_\chi$ by Lemma A.1.9, it follows that f_1 is really a surjection from ω to a. Hence we have $a = \{ f_1(n) : n \in \omega \}$. Let $u_n = f_1(n)$ for each $n \in \omega$. Since $\mathcal{H}_\chi \models$ "$\exists!x(x = f_1(n))$" and $\mathcal{H}_\chi \models$ "$u_n = f_1(n)$", it follows, by Lemma A.1.10, that $u_n \in M$ for every $n \in \omega$. Hence $x = \{ u_n : n \in \omega \} \subseteq M$. $\square$ (Lemma A.1.11)

In the following we shall give some applications of the ideas introduced above. As the first application, let us prove the following generalization of Theorem A.1.7 for $\kappa = \aleph_1$ using elementary substructures of $\mathcal{H}_\chi$ for some appropriate χ.

Theorem A.1.12 *If C is a closed unbounded subset of $[X]^{\aleph_0}$ and $Y \subseteq X$, then $\mathcal{D} = \{ u \cap Y : u \in C \}$ contains a closed unbounded subset of $[Y]^{\aleph_0}$.*

Proof Let χ be a regular cardinal such that $C, X \in \mathcal{H}_\chi$. First let us show that:

CLAIM 1. *If $M \prec \mathcal{H}_\chi$ is such that $C \in M$ and $|M| = \aleph_0$, then $X \cap M \in C$*

$\vdash$ Note that from $C \in M$, it follows that $X \in M$. Let $(x_n)_{n \in \omega}$ be an enumeration of $X \cap M$. Since $\mathcal{H}_\chi \models$ "C is cofinal in $[X]^{\aleph_0}$", we have also $M \models$ "C is cofinal in $[X]^{\aleph_0}$". Hence we can take inductively $u_n \in C \cap M$ such that $\bigcup_{m<n} u_m \cup \{ x_n \} \subseteq u_n$. (This is possible since we have $(u_m)_{m<n} \in M$ at the n'th step of the inductive construction.) By Lemma A.1.11, we have $u_n \subseteq M$ for every $n \in \omega$. Hence it follows that $\bigcup_{n \in \omega} u_n = X \cap M$. But since C is closed with respect to union of countable increasing chains, $\bigcup_{n<\omega} u_n \in C$ holds. $\dashv$ (CLAIM 1.)

Let

$$\mathcal{E} = \{ Y \cap M : M \prec \mathcal{H}_\chi, |M| = \aleph_0, C \in M \}.$$

By Theorem A.1.7, $\mathcal{E}$ contains a closed unbounded subset of $[Y]^{\aleph_0}$. Hence the following claim proves the theorem.

CLAIM 2. $\mathcal{E} \subseteq \mathcal{D}$.

$\vdash$ Let $M \prec \mathcal{H}_\chi$ be such that $|M| = \aleph_0$ and $C \in M$. By Claim 1, we have $X \cap M \in C$. Hence we have $Y \cap M = Y \cap (X \cap M) \in \mathcal{D}$. $\dashv$ (CLAIM 2.)
$\square$ (Theorem A.1.12)

Now we shall give a proof of the following general form of the Δ-*System Lemma* (see also 0.0.2):

Theorem A.1.13 *Assume that κ, χ are regular cardinals such that $\kappa < \chi$ and*

*) $|\alpha^{<\kappa}| < \chi$ *for every* $\alpha < \chi$.

If A is a set of cardinality $\geq \chi$ such that $|x| < \kappa$ for every $x \in A$, then there is $B \subseteq A$ of cardinality χ and a set r such that $x \cap y = r$ for any distinct $x, y \in B$.

For the proof of Theorem A.1.13, we use the following:

Lemma A.1.14 *If κ, χ are regular cardinals such that $\kappa < \chi$, *) as in Theorem A.1.13 holds and $x_0 \in \mathcal{H}_\chi$, then there is $M \prec \mathcal{H}_\chi$ such that $x_0 \in M$, $[M]^{<\kappa} \subseteq M$ and $M \cap \mathrm{Ord} \in \mathrm{Ord}$.*

Proof We construct inductively a continuously increasing chain $(M_\alpha)_{\alpha < \kappa}$ of elementary substructures of $\mathcal{H}_\chi$ such that

0) $x_0 \in M_0$;

1) $|M_\alpha| < \chi$ for every $\alpha < \kappa$;

2) $[M_\alpha]^{<\kappa} \subseteq M_{\alpha+1}$ and

3) $\sup(M_\alpha \cap \mathrm{Ord}) \subseteq M_{\alpha+1}$ for all $\alpha < \kappa$.

This is possible by Theorem A.1.4, a) and by the condition *). Let $M = \bigcup_{\alpha < \kappa} M_\alpha$. By Theorem A.1.4, b) , we have $M \prec \mathcal{H}_\chi$. By *0)*, $x_0 \in M$. By *1)* and since χ is regular we have $|M| < \chi$. By *2)* and since κ is regular we have $[M]^{<\kappa} \subseteq M$. By *3)*, $M \cap \mathrm{Ord}$ is an initial segment of Ord. Hence $M \cap \mathrm{Ord} \in \mathrm{Ord}$ holds. $\square$ (Lemma A.1.14)

Now we are ready to prove Theorem A.1.13.

Proof of Theorem A.1.13 Without loss of generality we may assume that $\bigcup A = \chi$. By Lemma A.1.14, there is $M \prec \mathcal{H}_\chi$ such that $A \in M$, $|M| < \chi$, $M \cap \mathrm{Ord} \in \mathrm{Ord}$ and $[M]^{<\kappa} \subseteq M$. Let $\alpha_0 = M \cap \mathrm{Ord}$ and let $u_0 \in A$ be such that $\sup u_0 > \alpha_0$. Let $r = u_0 \cap M$. Since $[M]^{<\kappa} \subseteq M$, we have $r \in M$.

CLAIM 1.

$M \models$ "for any ordinal α there is $u \in A$ such that $r \subseteq u$ and $\min(u \setminus r) > \alpha$".

$\vdash$ For an ordinal α in M, we have $\alpha < \alpha_0$. Since $\mathcal{H}_\chi \models$ "$r \subseteq u_0$ and $\min(u_0 \setminus r) > \alpha_0$", it follows that $\mathcal{H}_\chi \models$ "$\exists u \in A(r \subseteq u$ and $\min(u \setminus r) > \alpha)$". By $M \prec \mathcal{H}_\chi$ (and since $A, r, \alpha \in M$) it follows that $M \models$ "$\exists u \in A(r \subseteq u$ and

$\min(u \setminus r) > \alpha)$ ". $\dashv$ (CLAIM 1.)

By CLAIM 1. and $M \prec \mathcal{H}_\chi$, it follows that

$\mathcal{H}_\chi \models$ "for any ordinal α there is $u \in \mathcal{A}$ such that $r \subseteq u$ and $\min(u \setminus r) > \alpha$".

Hence we can take a sequence $(u_\alpha)_{\alpha<\chi}$ of elements of $\mathcal{A}$ inductively such that $r \subseteq u_\alpha$ and $\min(u_\alpha \setminus r) > \sup(\bigcup_{\beta<\alpha} u_\alpha)$ for every $\alpha < \chi$. Clearly $\{ u_\alpha : \alpha < \chi \}$ is as desired. $\square$ (Theorem A.1.13)

One advantage of the the proof of Theorem A.1.13 above is that almost the same proof also works for the following generalization of the theorem:

Theorem A.1.15 *Assume that κ, χ are regular cardinals such that $\kappa < \chi$ and*

*) $|\alpha^{<\kappa}| < \chi$ *for every* $\alpha < \chi$.

Let A be an algebraic structure. If $\mathcal{A}$ is a family of substructures such that $|\mathcal{A}| \geq \chi$ and $|B| < \kappa$ for every $B \in \mathcal{A}$, then there is $\mathcal{B} \subseteq \mathcal{A}$ of cardinality χ such that either $B \cap C = \emptyset$ for each two distinct $B, C \in \mathcal{B}$ or there is a substructure D of A such that $B \cap C = D$ holds for every two distinct $B, C \in \mathcal{B}$. $\square$

For the proof, we take χ and M so that in addition to the conditions in the proof of Theorem A.1.13, $A \in \mathcal{H}_\chi$ and $A \in M$ hold. Then it is enough to show that either $r = \emptyset$ or r is a substructure of A for r as in the proof. But this follows from:

Theorem A.1.16 *Let χ be a regular cardinal and B an algebraic structure. If $M \prec \mathcal{H}_\chi$ is such that $B \in M$ and A is a substructure of B then*

a) $B \cap M \neq \emptyset$ *and*

b) *either* $A \cap M = \emptyset$ *or* $A \cap M$ *is a substructure of* B.

Proof *a*): Since $\mathcal{H}_\chi \models$ "$B \neq \emptyset$" we have $M \models$ "$B \neq \emptyset$". Hence there is $b \in M$ such that $M \models$ "$b \in B$" *i.e.* $b \in B \cap M$.

b): Assume that $A \cap M \neq \emptyset$. We show that $A \cap M$ is closed with respect to the functions of B (here, we consider the constants of B as 0-place functions). Let $f = f(x_1, \dots, x_n)$ be a function of the algebraic structure B. Let $a_1, \dots, a_n \in A \cap M$ and $a_0 = f(a_1, \dots, a_n)$. Since $\mathcal{H}_\chi \models$ "$\exists! x \in A\, f(a_1, \dots, a_n) = x$" we have $a_0 \in M$ by Lemma A.1.10. On the other hand since A is closed with respect to f we have $a_0 \in A$. This shows that $A \cap M$ is closed with respect to f. $\square$ (Theorem A.1.16)

A.2 Characterizations of rc-filtered and σ-filtered Boolean algebras

In Chapter 2 we saw several characterizations of rc-filtered Boolean algebras (see Theorem 2.2.3). In this section we shall give some other characterizations of rc-filtered Boolean algebras. We shall also give some characterizations of σ-filtered Boolean algebras which will be used in section A.4.

Let us begin with the following characterization of rc-filtered Boolean algebras based on the notions introduced in the last section. For a Boolean algebra B, let us say that a cardinal χ is *cardinal!large enough* when χ is regular and $2^{|trcl(B)|}, 2^{2^{|trcl(B)|}}, \cdots < \chi$ holds (this definition of being "large enough" is made really large enough: in most cases $|trcl(B)| < \chi$ is enough). If it is clear which B is meant, we shall simply say that χ is large enough. Note that, if χ is large enough for B, then B, $P(B)$, $[B]^{\aleph_0}$, $^B B (= \{ f : f : B \to B \})$ and so on are elements and also subsets of $\mathcal{H}_\chi$. A part of the following characterization was used in a topological setting by I. Bandlow in [4] and [5].

Theorem A.2.1 *For a Boolean algebra B, the following are equivalent:*

1) B is rc-filtered;

2) For some, or equivalently any large enough χ, if $M \prec \mathcal{H}_\chi$ is countable and $B \in M$ then $B \cap M \leq_{rc} B$ holds;

3) For some, or equivalently any large enough χ, if $M \prec \mathcal{H}_\chi$ (not necessarily countable) is such that $B \in M$ then $B \cap M \leq_{rc} B$ holds.

Proof $1) \Rightarrow 3)$: Let χ be large enough for B and let $M \prec \mathcal{H}_\chi$ be such that $B \in M$. Suppose that B is rc-filtered then by definition there exist mappings $U, L : B \to [B]^{<\aleph_0}$ as in Definition 2.2.1 which witness (FN) of B. Since U, $L \in \mathcal{H}_\chi$ we have

$$\mathcal{H}_\chi \models \text{``}\exists L \exists U (U \text{ and } L \text{ witness (FN) of } B)\text{''}.$$

Since $M \prec \mathcal{H}_\chi$ and $B \in M$, it follows that

$$M \models \text{``}\exists L \exists U (U \text{ and } L \text{ witness (FN) of } B)\text{''}.$$

Hence there are $U_0, L_0 \in M$ such that

$$\mathcal{H}_\chi \models \text{``}U_0 \text{ and } L_0 \text{ witness (FN) of } B\text{''}.$$

Note that U_0 and L_0 are really witnesses of (FN) of B by absoluteness. Now for any $b \in B \cap M$, $U_0(b), L_0(b) \in M$ holds. Since $L_0(b)$ and $U_0(b)$ are finite, it follows by Lemma A.1.11 that $U_0(b), L_0(b) \subseteq M$. Thus $B \cap M$ is closed under mappings U_0 and L_0. By the Claim in the proof of Theorem 2.2.3, it follows that $B \cap M \leq_{rc} B$.

$3) \Rightarrow 2)$: Trivial.

$2) \Rightarrow 1)$: Let χ be as in $2)$. By Theorem A.1.7,

$$\{ B \cap M : M \prec \mathcal{H}_\chi, |M| = \aleph_0, B \in M \}$$

contains a closed unbounded subset of $[B]^{\aleph_0}$. Hence by Theorem 2.2.3 B is rc-filtered. □ (Theorem A.2.1)

Theorem A.2.1 very often allows us to provide a theorem on rc-filtered Boolean algebras with an elegant short proof. As an example, let us consider Corollary 2.3.6 which is restated here as Theorem A.2.2 below:

Theorem A.2.2 *If $A \leq_\sigma B$ and B is rc-filtered then A is also rc-filtered.*

Proof Let χ be large enough (for B). Let $f : B \to [A]^{\aleph_0}$ be such that for every $b \in B$, $f(b)$ generates the ideal $A \restriction b$. Then $f \in \mathcal{H}_\chi$. Let $M \prec \mathcal{H}_\chi$ be such that A, B, $f \in M$. By the proof of $2) \Rightarrow 1)$ of Theorem A.2.1, it is easily seen that it is enough to show that $A \cap M \leq_{rc} A$. This can be shown as follows: Let $a \in A$. Since $B \cap M \leq_{rc} B$ by Theorem A.2.1, there is $q_{B \cap M}^B(a) \in B \cap M$. By $f \in M$ and Lemma A.1.11, we have $f(q_{B \cap M}^B(a)) \subseteq A \cap M$. Since $f(q_{B \cap M}^B(a))$ generates $A \restriction q_{B \cap M}^B(a)$ and $a \leq q_{B \cap M}^B(a)$, there is $a' \in f(q_{B \cap M}^B(a))$ such that $a \leq a'$. Since $a' \in A \cap M \subseteq B \cap M$, it follows that $q_{B \cap M}^B(a) \leq a'$. Hence we have $a' = q_{B \cap M}^B(a)$. This shows that $q_{B \cap M}^B(a)$ is the projection of a onto $A \cap M$.
$\square$ (Theorem A.2.2)

For projective Boolean algebras we have the following analogue of Theorem A.2.1:

Lemma A.2.3 *Let B be a projective Boolean algebra and let χ be large enough for B. Then for any $M \prec \mathcal{H}_\chi$ such that $B \in M$, $B \cap M \leq_{rc} B$ holds and $B \cap M$ is projective.*

Proof $B \cap M \leq_{rc} B$ holds by Theorem A.2.1. By Theorem 1.3.2,(5), there exists a countable family $(A_j)_{j \in J}$ of subalgebras of B such that $\langle \bigcup_{j \in K} A_j \rangle \leq_{rc} B$ for every $K \subseteq J$. By $M \prec \mathcal{H}_\chi$ and $B \in M$ we may assume that $(A_j)_{j \in J} \in M$ holds. By Lemma A.1.11, we have $A_j \subseteq A \cap M$ for every $j \in J \cap M$ hence $(A_j)_{j \in (J \cap M)}$ is as in Theorem 1.3.2,(5) for $B \cap M$. This shows that $B \cap M$ is projective.
$\square$ (Lemma A.2.3)

Lemma A.2.3 actually characterizes projectivity because of the trivial reason: if $B \subseteq M$ then we have $B \cap M = B$. If restricted to $M \prec \mathcal{H}_\chi$ such that $|M| < |B|$, the assertion of Lemma A.2.3 does not characterize projectivity. This can be seen in the Boolean algebra constructed in Chapter 6.

Our next characterization of rc-filtered Boolean algebras uses an infinitary game played on Boolean algebras. For a Boolean algebra B, let $\mathcal{G}_{rc}^\omega(B)$ be the following game played by Players I and II: Players I and II choose countable subalgebras B_n, $n \in \omega$ of B in turn so that $B_0 \leq B_1 \leq B_2 \cdots$ holds. Thus a match in $\mathcal{G}_{rc}^\omega(B)$ looks like:

$$\begin{array}{llll} \text{Player I} & : & B_0, & B_2, & \cdots \\ \text{Player II} & : & B_1, & B_3, & \cdots \end{array}$$

Player II wins if $\bigcup_{n \in \omega} B_n \leq_{rc} B$ holds.

Theorem A.2.4 *For a Boolean algebra B, the following are equivalent:*

1) B *is rc-filtered;*

2) *Player II has a winning strategy in* $\mathcal{G}_{rc}^\omega(B)$.

Proof $1) \Rightarrow 2$): Let U and L be witnesses of (FN) of B (*i.e.* mappings as in Definition 2.2.1). By Theorem 2.2.3 and Corollary 2.2.7, Player II wins in $\mathcal{G}^{\omega}_{rc}(B)$ if she plays subalgebras of B closed with respect to U and L.

$2) \Rightarrow 1$): Assume that Player II has a winnings strategy in $\mathcal{G}^{\omega}_{rc}(B)$. Let χ be large enough for B. Let $M \prec \mathcal{H}_{\chi}$ be such that $B \in M$. Note that a strategy of Player II is a mapping from increasing sequences of countable subalgebras of B of odd length to countable subalgebras of B. Hence a winning strategy of Player II in $\mathcal{G}^{\omega}_{rc}(B)$ is an element of $\mathcal{H}_{\chi}$. Thus, by $M \prec \mathcal{H}_{\chi}$, there is a winning strategy $\sigma \in M$ of Player II in $\mathcal{G}^{\omega}_{rc}(B)$. Now by Theorem A.2.1, it is enough to show the following:

CLAIM 1. $B \cap M \leq_{rc} B$ holds.

$\vdash$ Let $B \cap M = \{ b_n : n \in \omega \}$. Let $(B_n)_{n \in \omega}$ be a match in $\mathcal{G}^{\omega}_{rc}(B)$ such that Player I chooses his move B_{2n} so that $B_{2n} \in M$ and $b_n \in B_{2n}$ hold and Player II plays according to σ. Such a match is possible since by $\sigma \in M$ we have also $\sigma((B_m)_{m \leq 2n}) \in M$ if $(B_m)_{m \leq 2n} \in M$. Now let $C = \bigcup_{n \in \omega} B_n$. By Lemma A.1.11, we have $B_n \subseteq M$ for every $n \in \omega$. Hence by the construction, we have $B \cap M = C$. But $C \leq_{rc} B$ since σ was a winning strategy for Player II.
$\dashv$ (CLAIM 1.) $\qquad\qquad\qquad\qquad$ $\square$ (Theorem A.2.4)

For σ-filtered Boolean algebras, we have characterizations similar to Theorem A.2.1 and Theorem A.2.4. If B is a σ-filtered Boolean algebra, let us call a mapping $I : B \to [B]^{\leq \aleph_0}$ as in Definition 4.2.1, a witness of (WFN) of B.

Proposition A.2.5 (Fuchino, Koppelberg, Shelah, [19]) *For a Boolean algebra B, the following are equivalent:*

1) B is σ-filtered;

2) For some, or equivalently, any large enough χ and for all $M \prec \mathcal{H}_{\chi}$, if $B \in M$ and $|M| = \aleph_1$ then $B \cap M \leq_{\sigma} B$ holds;

3) $\{ C \in [B]^{\aleph_1} : C \leq_{\sigma} B \}$ contains a closed unbounded subset of $[B]^{\aleph_1}$;

4) There exists a partial ordering $I = (I, \leq)$ and an indexed family $(B_i)_{i \in I}$ of subalgebras of B of cardinality $\aleph_1$ such that

i) $\{ B_i : i \in I \}$ is cofinal in $([B]^{\aleph_1}, \subseteq)$,

ii) I is directed and for any $i, j \in I$, if $i \leq j$ then $B_i \leq B_j$,

iii) for every well-ordered $I' \subseteq I$ of cofinality $\leq \omega_1$, $i^* = \sup I'$ exists and $B_{i^*} = \bigcup_{i \in I'} B_i$ holds, and

iv) $B_i \leq_{\sigma} B$ holds for every $i \in I$.

Proof $1) \Rightarrow 2$): Let f be a witness of (WFN) of B. Since χ is large enough for B, we have $B, f \in \mathcal{H}_{\chi}$. Let $M \prec H_{\chi}$ be such that $B \in M$ and $|M| = \aleph_1$. Then there is a witness f' of (WFN) of B in M. Clearly $B \cap M$ is closed with respect to f'. Hence by a similar argument to that for the Claim in the proof of Theorem 2.2.3, we can show that $B \cap M \leq_{\sigma} B$ holds.

$2) \Rightarrow 3$): By Theorem A.1.7.

$3) \Rightarrow 4$): Let $I \subseteq \{ C \in [B]^{\aleph_1} : C \leq_\sigma B \}$ be a closed unbounded subset of $[B]^{\aleph_1}$ with the subalgebra relation. For $A \in I$, let $B_A = A$. Then $(I, \leq)$ and $(B_A)_{A \in I}$ satisfy the conditions in 4).

$4) \Rightarrow 1$): we prove this in the following two claims. Let I and $(B_i)_{i \in I}$ be as in 4). For a directed $I' \subseteq I$ let $B_{I'} = \bigcup_{i \in I'} B_i$.

CLAIM 1. *If $I' \subseteq I$ is directed then $B_{I'} \leq_\sigma B$ holds.*

⊢ Otherwise there would be a $b \in B$ such that $B_{I'} \restriction b$ is not countably generated. Hence there exists an increasing sequence $(I_\alpha)_{\alpha < \omega_1}$ of countable directed subsets of I' such that $B_{I_\alpha} \restriction b$ does not generate $B_{I_{\alpha+1}} \restriction b$. By iii), $i_\alpha = \sup I_\alpha$ exists and $B_{i_\alpha} = B_{I_\alpha}$ for every $\alpha < \omega_1$. $(i_\alpha)_{\alpha < \omega_1}$ is an increasing sequence in I. Hence again by iii) there exists $i^* = \sup_{\alpha < \omega_1} i_\alpha$ and $B_{i^*} = \bigcup_{\alpha < \omega_1} B_{i_\alpha}$. By iv), $B_{i^*} \leq_\sigma B$. But by the construction, $B_{i^*} \restriction b$ is not countably generated. This is a contradiction. ⊣ (CLAIM 1.)

CLAIM 2. *If $I' \subseteq I$ is directed then $B_{I'}$ is σ-filtered.*

⊢ We prove the claim by induction on $|I'|$. If $|I'| \leq \aleph_1$, we have $|B_{I'}| = \aleph_1$. Hence by Corollary 4.2.6, $B_{I'}$ is σ-filtered. Assume that $|I'| = \kappa > \aleph_1$ and that we have proved the claim for every directed $I'' \subseteq I$ with $|I''| < \kappa$. By Iwamura's lemma there is a continuously increasing sequence $(I_\alpha)_{\alpha < \kappa}$ of directed subsets of I' such that $|I_\alpha| < \kappa$ for every $\alpha < \kappa$ and $I' = \bigcup_{\alpha < \kappa} I_\alpha$. Then $(B_{I_\alpha})_{\alpha < \kappa}$ is a continuously increasing sequence of subalgebras of $B_{I'}$ and $B_{I'} = \bigcup_{\alpha < \kappa} B_{I_\alpha}$. By the induction hypothesis, B_{I_α} is σ-filtered and, by CLAIM 1., we have $B_{I_\alpha} \leq_\sigma B$. Hence by Lemma 4.2.4, $B_{I'}$ is also σ-filtered. ⊣ (CLAIM 2.)

Now by applying CLAIM 2. to $I' = I$, we can conclude that $B = B_I$ is σ-filtered. □ (Proposition A.2.5)

We shall give yet another characterization of σ-filtered Boolean algebras by means of the following game. For a Boolean algebra B let $\mathcal{G}_\sigma^{\omega_1}(B)$ the following game played by Players I and II: in a match in $\mathcal{G}_\sigma^{\omega_1}(B)$, Players I and II choose countable subsets X_α and Y_α alternately for $\alpha < \omega_1$ so that

$$X_0 \subseteq Y_0 \subseteq X_1 \subseteq Y_1 \subseteq \cdots \subseteq X_\alpha \subseteq Y_\alpha \subseteq \cdots \subseteq X_\beta \subseteq Y_\beta \subseteq \cdots$$

holds for $\alpha \leq \beta < \omega_1$. So a match in $\mathcal{G}_\sigma^{\omega_1}(B)$ looks like

$$\begin{array}{llllll} Player\ I & : & X_0, & X_1, & \ldots, & X_\alpha, \ldots \\ Player\ II & : & Y_0, & Y_1, & \ldots, & Y_\alpha, \ \ldots \end{array}$$

where $\alpha < \omega_1$. Player II wins the match if $\bigcup_{\alpha < \omega_1} X_\alpha = \bigcup_{\alpha < \omega_1} Y_\alpha$ is a σ-subalgebra of B. Let us call a strategy τ of Player II simple if, in τ, each Y_α is decided from the information of the set $X_\alpha \subseteq B$ alone (*i.e.* also independent of α).

For a large enough χ (with respect to B), an elementary substructure M of $\mathcal{H}_\chi$ is said to be V_{ω_1}-*like over* B if there is an increasing sequence $(M_\alpha)_{\alpha < \omega_1}$

of countable elementary substructures of M such that $B \in M_0$, $M_\alpha \in M_{\alpha+1}$ holds for all $\alpha < \omega_1$ and $M = \bigcup_{\alpha < \omega_1} B_\alpha$. If M is V_{ω_1}-like over B, we say that a sequence $(M_\alpha)_{\alpha < \omega_1}$ as above witnesses the V_{ω_1}-likeness of M over B.

Lemma A.2.6 *Suppose $B \in \mathcal{H}_\chi$ for some uncountable χ. For any $X \in [\mathcal{H}_\chi]^{\leq \aleph_1}$ there exists a V_{ω_1}-like elementary substructure M of $\mathcal{H}_\kappa$ over B such that $X \subseteq M$.*

Proof Let $X = \{ x_\alpha : \alpha < \omega_1 \}$. By Theorem A.1.4, we can construct a continuously increasing sequence $(M_\alpha)_{\alpha < \omega_1}$ of countable elementary substructures of $\mathcal{H}_\chi$ such that $B \in M_0$ and $M_\alpha, x_\alpha \in M_{\alpha+1}$ for every $\alpha < \omega_1$. $M = \bigcup_{\alpha < \omega_1} M_\alpha$ is then as desired. $\qquad\qquad\qquad\qquad\qquad\qquad\qquad\qquad\qquad$ $\square$ (Lemma A.2.6)

Proposition A.2.7 (Fuchino, Koppelberg, Shelah [19]) *For a Boolean algebra B, the following are equivalent:*

1) B *is σ-filtered;*

2) *Player II has a simple winning strategy in $\mathcal{G}_\sigma^{\omega_1}(B)$;*

3) *For any large enough χ and $M \prec \mathcal{H}_\chi$, if M is V_{ω_1}-like over B then $B \cap M \leq_\sigma B$ holds.*

Proof *1)* $\Rightarrow$ *2)*: Let $f : B \to [B]^{\aleph_0}$ be a witness of (WFN) of B. Then Player II can win by the following strategy: In α'th move, Player II chooses Y_α so that $X_\alpha \subseteq Y_\alpha$ and Y_α is a countable subalgebra of B closed under f. After ω_1 moves, $\bigcup_{\alpha < \omega_1} Y_\alpha$ will be a subalgebra of B closed under f. Hence it will be a σ-subalgebra of B.

2) $\Rightarrow$ *3)*: Let M be a V_{ω_1}-like elementary substructure of $\mathcal{H}_\chi$ over B. We have to show that $B \cap M \leq_\sigma B$ holds. Let $(M_\alpha)_{\alpha < \omega}$ witnesses the V_{ω_1}-likeness of M over B. By $B \in M_0$ and $M_0 \prec \mathcal{H}_\chi$, there is a winning strategy $\tau \in M_0$ for Player II in $\mathcal{G}_\sigma^{\omega_1}(B)$. (Hence $\tau \in M_\alpha$ holds for every $\alpha < \omega_1$.) Let $(X_\alpha, Y_\alpha)_{\alpha < \omega_1}$ be such a match in $\mathcal{G}_\sigma^{\omega_1}(B)$ that at his α'th move, Player I played $B \cap M_\alpha$ and Player II played always according to τ. Such a game is possible since if Player I has chosen $B \cap M_\alpha$ at his α'th move, then $B \cap M_\alpha \in M_{\alpha+1}$. Hence Player II's move Y_α taken according to τ is also an element of $M_{\alpha+1}$. Since Y_α is countable, it follows that $Y_\alpha \subseteq M_{\alpha+1}$. Hence Player I may take $B \cap M_{\alpha+1}$ at his next move.

Now we have $B \cap M = B \cap (\bigcup_{\alpha < \omega_1} M_\alpha) \leq_\sigma B$ since τ was a winning strategy of Player II.

3) $\Rightarrow$ *1)*: First note the following:

CLAIM 1. *If M is a V_{ω_1}-like elementary substructure of $\mathcal{H}_\chi$ over B then $\omega_1 \subseteq M$. Hence if x is of cardinality $\aleph_1$ and $x \in M$ then we have $x \subseteq M$.*

$\vdash$ Let $(M_\alpha)_{\alpha < \omega_1}$ witness the V_{ω_1}-likeness of M over B. Assume that $\omega_1 \not\subseteq M$. Let

$$\alpha_0 = \min\{ \alpha \in \omega_1 : \alpha \notin M \}.$$

Then we have $\alpha_0 \subseteq M$. Let

$$\alpha_1 = \min\{ \alpha \leq \omega_1 : \alpha_0 \leq \alpha, \alpha \in M \}.$$

Since α_0 is countable there exists $\alpha < \omega_1$ such that $\alpha_0 \subseteq M_\alpha$ and $\alpha_1 \in M_\alpha$ hold. Since $M_\alpha \in M_{\alpha+1}$, $\alpha_0 = \{\beta \in M_\alpha : \beta < \alpha_1\}$ is an element of $M_{\alpha+1} \subseteq M$. This is a contradiction. Hence $\omega_1 \subseteq M$.

Now, if x is of cardinality $\aleph_1$ and $x \in M$ then there is a surjection $f : \omega_1 \to x$ in M. Since $\omega_1 \subseteq M$, it follows that $x \subseteq M$. $\dashv$ (CLAIM 1.)

Now let

$$I = \{ M : \quad |M| = \aleph_1, \, M \text{ is a union of an increasing sequence}$$
$$\text{of } V_{\omega_1}\text{-like elementary substructures of } \mathcal{H}_\chi \text{ over } B \}.$$

Note that, in the definition of I, we do not demand that the sequence of V_{ω_1}-like substructures of $\mathcal{H}_\chi$ is strictly increasing. Hence every V_{ω_1}-like substructure of $\mathcal{H}_\chi$ over B is also an element of I. For M, $N \in I$, let $M \leq N \Leftrightarrow M \in N$. Note also that, by CLAIM 1., we have $M \subseteq N$ if $M \leq N$. For $M \in I$, let $B_M = B \cap M$. We show that I and $(B_M)_{M \in I}$ satisfy $i) - iv)$ of Proposition A.2.5, 4). From this it follows, by Proposition A.2.5, that B is σ-filtered.

$i)$ $\{ B_M : M \in I \}$ *is cofinal in* $([B]^{\aleph_1}, \subseteq)$:

Let $X \in [B]^{\aleph_1}$, say $X = \{ b_\alpha : \alpha < \omega_1 \}$. By Theorem A.1.4, $a)$, we can construct an increasing sequence $(M_\alpha)_{\alpha < \omega_1}$ of countable elementary substructures of $\mathcal{H}_\chi$ inductively so that M_α, $b_\alpha \in M_{\alpha+1}$ for every $\alpha < \omega_1$. Then $M = \bigcup_{\alpha < \omega_1} M_\alpha$ is V_{ω_1}-like over B and $X \subseteq M$. Hence we have $X \subseteq B_M$

$ii)$ I *is directed and for any* M, $N \in I$, *if* $M \leq N$ *then* $B_M \subseteq B_N$:

The directedness of I can be proved as in $i)$. The second half of the assertion is trivial by the definition of B_M.

$iii)$ *for every well-ordered* $I' \subseteq I$ *of cofinality* $\leq \omega_1$, $i^* = \sup I'$ *exists and* $B_{i^*} = \bigcup_{i \in I'} B_i$:

Let $(M_\alpha)_{\alpha < \delta}$ be an increasing sequence of elements of I for $\delta \leq \omega_1$. For each $\alpha < \delta$, let $(M_{\alpha,\beta})_{\beta < \eta_\alpha}$ be an increasing sequence of V_{ω_1}-like elementary substructures of $\mathcal{H}_\chi$ over B such that $M_\alpha = \bigcup_{\beta < \eta_\alpha} M_{\alpha,\beta}$. Since $M_\alpha \in M_{\alpha+1}$ holds for every $\alpha < \delta$ there is $\beta_\alpha < \eta_{\alpha+1}$ such that $M_\alpha \in M_{\alpha+1,\beta_\alpha}$ holds. By CLAIM 1., it follows that $M_\alpha \subseteq M_{\alpha+1,\beta_\alpha}$. Hence $\bigcup_{\alpha < \delta} M_\alpha = \bigcup_{\alpha < \delta} M_{\alpha+1,\beta_\alpha}$. This shows that $\bigcup_{\alpha < \delta} M_\alpha \in I$.

$iv)$ $B_M \leq_\sigma B$ *for every* $M \in I$:

If $M \in I$ is union of a ω-chain $(M_n)_{n < \omega}$ of V_{ω_1}-like elementary substructures of $\mathcal{H}_\chi$ over B, then we have $B \cap M_n \leq_\sigma B$ for every $n \in \omega$ by 3). Hence we have $B \cap M = \bigcup_{n < \omega} (B \cap M_n) \leq_\sigma B$. If $M \in I$ is union of ω_1-chain of V_{ω_1}-like elementary substructures of $\mathcal{H}_\chi$ over B, then if follows from the next claim that M itself is V_{ω_1}-like over B. Hence we have $B \cap M \leq_\sigma B$ by 3).

CLAIM 2. *If* $(N_\alpha)_{\alpha < \omega_1}$ *is an increasing sequence of* V_{ω_1}-*like elementary substructures of* $\mathcal{H}_\chi$ *over* B, *then* $M = \bigcup_{\alpha < \omega_1} N_\alpha$ *is also* V_{ω_1}-*like elementary substructure of* $\mathcal{H}_\chi$ *over* B.

⊢ It is clear that M is an elementary substructure of $\mathcal{H}_\chi$. To prove that M is V_{ω_1}-like over B, let $M = \{\, m_\xi \,:\, \xi \in \omega_1 \,\}$ and, for each $\alpha < \omega_1$, let $(N_{\alpha,\beta})_{\beta<\omega_1}$ be an increasing sequence of countable elementary substructures of $\mathcal{H}_\chi$ witnessing the V_{ω_1}-likeness of N_α. Since $N_{\alpha,\beta} \in N_\alpha \subseteq M$ and $\bigcup_{\beta<\omega_1} N_{\alpha,\beta} = N_\alpha$, we can choose $\alpha_\xi, \beta_\xi < \omega_1$ for $\xi < \omega_1$ inductively such that

a) $(N_{\alpha_\xi,\beta_\xi})_{\xi<\omega_1}$ is an increasing sequence,

b) $N_{\alpha_\xi,\beta_\xi} \in N_{\alpha_{\xi+1},\beta_{\xi+1}}$ holds for every $\xi < \omega_1$ and

c) $m_\xi \in N_{\alpha_\xi,\beta_\xi}$ for every $\xi < \omega_1$.

Clearly $(N_{\alpha_\xi,\beta_\xi})_{\xi<\omega_1}$ witnesses the V_{ω_1}-likeness of M over B. ⊣ (CLAIM 2.)

□ (Proposition A.2.7)

A.3 Answers to some other questions of Ščepin under $V = L$

In section 2.10, we saw solutions to Questions 3 and 9 of Ščepin in [55]. In this and the following section, we shall give solutions to two more questions from [55], namely Questions 7 and 8. To formulate these questions let us introduce the following notion:

For a regular cardinal κ, we say that a Boolean algebra B is *κ-projectively filtered* if there is a filtration $(B_i)_{i \in I}$ of B such that

1) B_i is projective for every $i \in I$ and

2)$_\kappa$ every chain $(i_\alpha)_{\alpha<\delta}$ in I of length $\delta < \kappa$ has supremum i^*.

Note that, by the definition of filtration, it follows form 2)$_\kappa$ that $B_{i^*} = \bigcup_{\alpha<\delta} B_{i_\alpha}$ holds. Also note that we only require here that each B_i is a subalgebra of B: B_i need not to be, for example, relatively complete subalgebra of B. Also there is no restriction on the cardinality of B_i. Let us call a filtration $(B_i)_{i \in I}$ of B with the properties 1) and 2)$_\kappa$ a *κ-projective filtration* of B.

Lemma A.3.1 *Let B be a Boolean algebra. Then*

a) *If B is rc-filtered then B is $\aleph_2$-projectively filtered.*

b) *If B is projective then B is κ-projectively filtered for every $\kappa \in \mathrm{Card}$.*

c) *Every Boolean algebra is $\aleph_1$-projectively filtered. In particular, there is a non rc-filtered Boolean algebra B which is $\aleph_1$-projectively filtered.*

d) *There is a rc-filtered Boolean algebra B which is not $\aleph_3$-projectively filtered.*

Proof *a)*: Assume that B is rc-filtered. Let χ be large enough for B and let

$$I = \{\, M \,:\, M \prec \mathcal{H}_\chi, \, |M| = \aleph_1, \, B \in M \,\}$$

with the ordering: $M \leq N \Leftrightarrow M \subseteq M$. For $M \in I$, let $B_M = B \cap M$. By Theorem A.2.1, $B_M \leq_{\mathrm{rc}} B$ for every $M \in I$. Hence by Corollary 2.2.7, B_M, $M \in I$ are projective. By Theorem A.1.4, *b)*, $(I, \leq)$ also satisfies 2)$_{\aleph_2}$ in the definition of $\aleph_2$-projectively filtered Boolean algebras. Thus $(B_M)_{M \in I}$ is a $\aleph_2$-projective filtration of B.

b): If B is projective then, letting $I = \{\emptyset\}$ and $B_\emptyset = B$, $(B_i)_{i \in I}$ is a κ-projective filtration for every regular κ.

c): Let $I = \{C \leq B : C \text{ is countable}\}$. Regard I as $(I, \leq)$ where $\leq$ is just the subalgebra relation. For $C \in I$, let $B_C = C$. Then $(B_C)_{C \in I}$ a $\aleph_1$-projective filtration of B.

d): The Boolean algebra considered in Section 6.3 has the desired property.
$$\square \text{ (Lemma A.3.1)}$$

Questions 7 and 8 in Ščepin [55] ask if the converse of Lemma A.3.1, *a*), *b*) also holds.

QUESTION 7. *Is any $\aleph_2$-projectively filtered Boolean algebra rc-filtered?*

QUESTION 8. *Is any $\aleph_3$-projectively filtered Boolean algebra projective?*

Lemma A.3.1, *c*) and *d*) show that we would have obtained negative answers to Questions 7 and 8, if $\aleph_2$ and $\aleph_3$ in the questions were replaced by smaller cardinals. In this and the following section we show that the answer to Question 7 is independent from set theory (see Theorem A.3.2 and Theorem A.4.9 below). Theorem A.3.2 below also shows that, under $V = L$, there is no cardinal κ such that κ-projective filtration characterizes projective Boolean algebras. In particular, under $V = L$, the answer to Question 8 is negative. Actually, by a result of S. Shelah in [20], the answer to Question 8 is even negative in ZFC alone.

If κ is a supercompact cardinal then κ-projective filtration characterizes projective Boolean algebras. From the consistency of ZFC + existence of a supercompact cardinal, we can prove the consistency of the assertion "$2^{\aleph_0}$-projective filtration characterizes the projectivity" (see [16]).

Theorem A.3.2 ([16]) $(V = L)$ *For every regular non weakly-compact κ there is a Boolean algebra B of cardinality κ such that B is κ-projectively filtered but not rc-filtered (hence, in particular, not projective).*

The construction of the Boolean algebra B in Theorem A.3.2 is similar to that of the Boolean algebra given in Proposition 2.10.1. $V = L$ is needed here so that the following principle E_ω^κ holds for every non-weakly compact regular κ:

E_ω^κ :   *There exists $S \subseteq \{\alpha < \kappa : \mathrm{cf}(\alpha) = \omega\}$ such that S is stationary but $S \cap \alpha$ is non-stationary for every limit ordinal $\alpha < \kappa$.*

For the proof of the following theorem see e.g. [8].

Theorem A.3.3 $(V = L)$ E_ω^κ *holds for every non-weakly-compact regular κ.* $\square$

For the construction of B in Theorem A.3.2 we also need the following:

Lemma A.3.4 *Let* $(B_n)_{n \in \omega}$ *be a strictly increasing sequence of free Boolean algebras such that* $B_n \leq_{free} B_{n+1}$ *for every* $n \in \omega$ *and let* $B = \bigcup_{n \in \omega} B_n$. *Then there exists a free Boolean algebra* C *such that* $B \leq C$, B *is not relatively complete in* C *but* $B_n \leq_{free} C$ *for every* $n \in \omega$.

Proof The idea of the following proof already appeared in the proof of Proposition 2.10.1. Let B^* and C^* be countable atomless Boolean algebras (hence $B^* \cong C^* \cong \mathrm{Fr}\omega$) such that $B^* \leq C^*$ but B^* is not relatively complete in C^*. Without loss of generality we may assume that $B^* = \mathrm{Fr}\omega$. Let $B_n^* = \mathrm{Fr}(\{0, \ldots, n-1\})$ for every $n \in \omega$ (so $B_n^* \leq B^*$). Since B_n^* is finite, we have $B_n^* \leq_{rc} C^*$. Hence by Sirota's Lemma (Lemma 1.4.10) $B_n^* \leq_{free} C^*$ holds for every $n \in \omega$. We can find an increasing sequence $(X_n)_{n \in \omega}$ of sets such that, letting $B_n' = B_n^* \otimes \mathrm{Fr}(X_n)$ and $B' = B^* \otimes \mathrm{Fr}(\bigcup_{n \in \omega} X_n)$, there exists an isomorphism f from B' to B such that $f[B_n'] = B_n$ holds for every $n \in \omega$. Let C be a Boolean algebra extending B which corresponds $C' = C^* \otimes \mathrm{Fr}(\bigcup_{n \in \omega} X_n)$ over f. Then C is as desired.   □ (Lemma A.3.4)

Now we are ready to prove Theorem A.3.2.

Proof of Theorem A.3.2   Let κ be regular and non-weakly-compact and let $S \subseteq \{\alpha < \kappa : \mathrm{cf}(\alpha) = \omega\}$ be as in the definition of E_ω^κ. We construct inductively a continuously strictly increasing sequence $(B_\alpha)_{\alpha < \kappa}$ of free Boolean algebras such that for every $\alpha < \beta < \kappa$,

0) $|B_\alpha| = |\alpha|$,

1) $B_\alpha \leq_{free} B_\beta$ if $\alpha \notin S$ and

2) B_α is not a relatively complete subalgebra of B_β if $\alpha \in S$.

Assume that we have constructed such $(B_\alpha)_{\alpha < \kappa}$. Let $B = \bigcup_{\alpha < \kappa} B_\alpha$. Clearly $(B_\alpha)_{\alpha < \kappa}$ is a κ-projective filtration of B. Hence B is κ-projectively filtered. On the other hand, if $\alpha \in S$, B_α is not a relatively complete subalgebra of B by *2)*. Hence

$$\{A : A \leq B \,|A| < \kappa, A \text{ is not a relatively complete subalgebra of } B\}$$

is a stationary subset[8] of $[B]^{<\kappa}$. Hence by Theorem A.1.7 and Theorem A.2.1, it follows that B is not rc-filtered.

For the construction of $(B_\alpha)_{\alpha < \kappa}$ let us assume that we have already constructed a continuously strictly increasing sequence $(B_\alpha)_{\alpha < \gamma}$ for some $\gamma < \kappa$ satisfying *0)*, *1)*, *2)*. We show that we can take B_γ so that $(B_\alpha)_{\alpha \leq \gamma}$ also satisfies *0)*, *1)*, *2)*.

If γ is a limit $B_\gamma = \bigcup_{\alpha < \gamma} B_\alpha$ will do. To see that B_γ is free, recall that $S \cap \gamma$ is non-stationary in γ. Hence there is a continuously increasing sequence $(\beta_\alpha)_{\alpha < \xi}$ of ordinals $< \gamma$ such that $\beta_\alpha \notin S$ for every $\alpha < \xi$ and $\gamma = \sup_{\alpha < \xi} \beta_\alpha$. By *1)*, we have $B_{\beta_\alpha} \leq_{free} B_{\beta_{\alpha'}}$ for every $\alpha < \alpha' < \xi$ and each B_{β_α}, $\alpha < \xi$ is free. Hence it follows that $B_\gamma = \bigcup_{\alpha < \xi} B_{\beta_\alpha}$ is free.

---

[8] $S \subseteq [X]^{<\kappa}$ is said to be stationary if S is not disjoint to any closed unbounded subset of $[X]^{<\kappa}$.

If $\gamma = \delta + 1$ and $\delta \notin S$ then let $B_\gamma = B_\delta \otimes \mathrm{Fr}\,\omega$. Then it is easily seen that $(B_\alpha)_{\alpha \leq \gamma}$ satisfies $0)$, $1)$, $2)$.

If $\gamma = \delta + 1$ and $\delta \in S$, we have $\mathrm{cf}(\delta) = \omega$. Hence we can find a strictly increasing sequence of ordinals $(\alpha_n)_{n \in \omega}$ such that $\alpha_n \notin S$ for all $n \in \omega$ and $\delta = \sup_{n \in \omega} \alpha_n$. By $1)$, we have $B_{\alpha_n} \leq_{free} B_{\alpha_m}$ for $n < m < \omega$ and $B_\delta = \bigcup_{n \in \omega} B_{\alpha_n}$. Hence by Lemma A.3.4, there is a free Boolean algebra $B_\gamma \geq B_\delta$ such that $B_{\alpha_n} \leq_{free} B_\gamma$ for every $n \in \omega$ but B_δ is not a relatively complete subalgebra of B_γ. $(B_\alpha)_{\alpha \leq \gamma}$ satisfies $0)$, $1)$, $2)$. To see that $1)$ is satisfied for $\beta = \gamma$, take an arbitrary $\alpha < \gamma$ such that $\alpha \notin S$. Then there is $n \in \omega$ such that $\alpha \leq \alpha_n$ holds. Since $B_\alpha \leq_{free} B_{\alpha_n}$ by $1)$ for $(B_\alpha)_{\alpha < \gamma}$ and since $B_{\alpha_n} \leq_{free} B_\gamma$, if follows that $B_\alpha \leq_{free} B_\gamma$. □ (Theorem A.3.2)

A.4 $\aleph_2$-projectively filtered Boolean algebras

In this section, we shall study further the class of $\aleph_2$-projectively filtered Boolean algebras introduced in the last section. We shall give a characterization of these Boolean algebras (Theorem A.4.7) in the spirit of Theorem A.2.1. Then, in Theorem A.4.9, we shall show the consistency of the statement that the class of $\aleph_2$-projectively filtered Boolean algebras is exactly the class of uncountable rc-filtered Boolean algebras. Hence, together with Theorem A.3.2, we obtain the independence of Ščepin's Question 7 in [55] (in the Boolean algebraic reformulation given in the last section). Let us begin with some lemmas we need for these results.

Lemma A.4.1 *If B is $\aleph_2$-projectively filtered then B satisfies the ccc.*

Proof Otherwise there would be a pairwise disjoint $X \subseteq B$ of cardinality $\aleph_1$. If $(B_i)_{i \in I}$ is a $\aleph_2$-projective filtration of B, there is $i^* \in I$ such that $X \subseteq B_{i^*}$. But this is impossible since B_{i^*} is projective by the definition of $\aleph_2$-projective filtration and hence satisfies the ccc. □ (Lemma A.4.1)

Lemma A.4.2 *Suppose that B is $\aleph_2$-projectively filtered and $(B_i)_{i \in I}$ is a $\aleph_2$-projective filtration of B. If χ is large enough and M is a V_{ω_1}-like elementary substructure of $\mathcal{H}_\chi$ over B such that $(B_i)_{i \in I} \in M$, then $B \cap M$ is a projective Boolean algebra and $B \cap M \leq_{rc} B_{i^*}$ holds for $i^* = \sup I \cap M$.*

Proof Let $(M_\alpha)_{\alpha < \omega_1}$ be an increasing sequence of countable elementary substructures of M witnessing the V_{ω_1}-likeness of M. Without loss of generality, we may assume that $(B_i)_{i \in I}$ is an element of M_0. For $\alpha < \omega_1$, let $i_\alpha = \sup I \cap M_\alpha$. i_α exists since $I \cap M_\alpha$ is countable and directed by $(I, \leq) \in M_\alpha$. By $i_\alpha \in M_{\alpha+1} \subseteq M$, we have $B_{i_\alpha} \in M$. Since B_{i_α} is projective, we have $B_{i_\alpha} \cap M \leq_{rc} B_{i_\alpha}$ by Theorem A.2.1. Clearly $i^* = \sup\{ i_\alpha : \alpha < \omega_1 \}$. Hence we have $B \cap M = \bigcup_{\alpha < \omega_1} (B_{i_\alpha} \cap M) = B_{i^*} \cap M$.

CLAIM 1. $B_{i^*} \cap M \leq_{rc} B_{i^*}$.

⊢ Otherwise there would be $b \in B_{i^*}$ such that $p^{B_{i^*}}_{B_{i^*} \cap M}(b)$ does not exist. Without loss of generality, we may assume that $b \in B_{i_0}$. Then $(p^{B_{i_\alpha}}_{B_{i_\alpha} \cap M}(b))_{\alpha < \omega_1}$ is not an eventually constant increasing sequence of elements of B_{i^*}. But this is a contradiction, since B_{i^*} satisfies the ccc. $\dashv$ (CLAIM 1.)

By Corollary 2.2.7 and Theorem 2.3.1, it follows that $B_{i^*} \cap M$ is projective.
$\square$ (Lemma A.4.2)

Lemma A.4.3 *If B is $\aleph_2$-projectively filtered then B satisfies the Bockstein separation property.*

Proof Let $(B_i)_{i \in I}$ be a $\aleph_2$-projective filtration of B and let χ be large enough such that $(B_i)_{i \in I} \in \mathcal{H}_\chi$. Suppose the there is a regular ideal J on B which is not countably generated. Similarly to the proof of Lemma A.2.6, we can find a V_{ω_1}-like elementary substructure M of $\mathcal{H}_\chi$ over B such that $(B_i)_{i \in I}, J \in M$ and $J \cap M$ is not countably generated as an ideal on $B \cap M$. Since $M \models$ "J is a regular ideal", $J \cap M$ is a regular ideal on $B \cap M$. But, since $B \cap M$ is projective and hence satisfies the Bockstein separation property by Proposition 2.3.3, it follows that $J \cap M$ is countably generated. This is a contradiction.
$\square$ (Lemma A.4.3)

For Boolean algebras B, C such that $C \leq B$ and $b \in B$, let

$$tp_C(b) = (\{ c \in C : c \leq b \}, \{ c \in C : c \wedge b = 0 \}).$$

For those familiar with model theory, we note that $tp_C(b)$ almost decides the model theoretic type of b over C; if B and C are atomless, $tp_C(b)$ actually decides the model theoretic type of b over C.

Lemma A.4.4 *If B is a Boolean algebra such that $\{ A \leq B : |A| = \aleph_1, A$ is projective$\}$ is cofinal in $([B]^{\aleph_1}, \subseteq)$ then for any countable $C \leq B$, the set $\{ tp_C(b) : b \in B \}$ is countable.*

Proof Assume that $\{ tp_C(b) : b \in B \}$ were uncountable. Then we can take elements b_α, $\alpha < \omega_1$ of B such that $tp_C(b_\alpha)$, $\alpha < \omega_1$ are pairwise distinct. By the assumption, there is $A \leq B$ such that A is projective and $C, \{ b_\alpha : \alpha < \omega_1 \} \subseteq A$. Let C' be a countable subalgebra of B such that $C \leq C'$ and $C' \leq_{proj} A$. Without loss of generality, we may assume that $b_\alpha \in A \setminus C'$ for every $\alpha < \omega_1$. Clearly $\{ tp_{C'}(a) : a \in A \}$ is countable. Hence there are $\alpha < \beta < \omega_1$ such that $tp_{C'}(b_\alpha) = tp_{C'}(b_\beta)$. It follows that $tp_C(b_\alpha) = tp_C(b_\beta)$. This is a contradiction.
$\square$ (Lemma A.4.4)

Note that, if B is $\aleph_2$-projectively filtered, then B satisfies the condition in Lemma A.4.4. Let us call a Boolean algebra B stable if, for every countable $C \leq B$, the set $\{ tp_C(b) : b \in B \}$ is countable.

Proposition A.4.5 *Suppose that B is stable and, has the ccc and the Bockstein separation property. If χ is large enough and M is a V_{ω_1}-like elementary substructure of $\mathcal{H}_\chi$ over B then $B \cap M \leq_{rc} B$.*

Proof Let $(M_\alpha)_{\alpha < \omega_1}$ be a witness of V_{ω_1}-likeness of M over B. Take an arbitrary $b \in B$. We show that $q^B_{B \cap M}(b)$ exists. Let U be a maximal pairwise disjoint subset of $B \cap M \restriction -b$. Since B satisfies the ccc, U is countable. Hence there is $\alpha^* < \omega_1$ such that $U \subseteq M_{\alpha^*}$. Now by Lemma A.4.4 and Lemma A.1.11, there is $b_0 \in M_{\alpha^*+1}$ such that $tp_{B \cap M_{\alpha^*}}(b) = tp_{B \cap M_{\alpha^*}}(b_0)$. Let

$$t = \{ c \in B \cap M_{\alpha^*} : c \wedge b_0 = 0 \}.$$

Then we have $t \in M_{\alpha^*+1}$ and $U \subseteq t$. Let

$$J = t^d = \{ d \in B : d \wedge c = 0 \text{ for every } c \in t \}.$$

Then J is a regular ideal on B and $J \in M_{\alpha^*+1}$. Since B has the Bockstein separation property, there exists a countable $X \in M_{\alpha^*+1}$ generating J. By Lemma A.1.11, we have $X \subseteq M_{\alpha^*+1}$. Hence $X \subseteq B \cap M$. Now, since $b \in J$, there is $d \in X$ such that $b \le d$. We show that d is the projection of b onto $B \cap M$. Otherwise there would be $c \in B \cap M$ such that $b \le c$ and $d \wedge -c \ne 0$. Since $d \wedge -c \in B \cap M$, by maximality of U and $(d \wedge -c) \wedge b = 0$, we have $(d \wedge -c) \wedge e \ne 0$ for some $e \in U$. But this is a contradiction to $d \in J \,(= t^d \subseteq U^d)$.

$\Box$ (Proposition A.4.5)

Corollary A.4.6 *If B is stable and, has the ccc and the Bockstein separation property, then B is σ-filtered. In particular, if B is $\aleph_2$-projectively filtered then B is σ-filtered.*

Proof The first assertion follows from Propositions A.2.7 and A.4.5. For the second, note that by Lemmas A.4.1, A.4.3, A.4.4, a $\aleph_2$-projectively filtered Boolean algebra is stable and, has the ccc and the Bockstein separation property.

$\Box$ (Corollary A.4.6)

The $\aleph_2$-projectively filtered Boolean algebras also enjoy a characterization similar to Theorem A.2.1:

Theorem A.4.7 *For a Boolean algebra B the following are equivalent:*

1) B is $\aleph_2$-projectively filtered;

2) For some, or equivalently any large enough χ, if $M \prec \mathcal{H}_\chi$ is V_{ω_1}-like over B then $B \cap M$ is projective and $B \cap M \le_{rc} B$;

3) For some, or equivalently any large enough χ, if $M \prec \mathcal{H}_\chi$ is such that $B \in M$ and $|M| = \aleph_1$ then $B \cap M$ is projective.

Proof *1)* $\Rightarrow$ *2)* follows from Lemmas A.4.2, A.4.3, A.4.4 and Proposition A.4.5. *2)* $\Rightarrow$ *3)*: By Proposition A.2.7, B is σ-filtered. Hence by Proposition A.2.5, $B \cap M \le_\sigma B$ holds for any $M \prec \mathcal{H}_\chi$ such that $B \in M$ and $|M| = \aleph_1$. By Lemma A.2.6, there is a V_{ω_1}-like $N \prec \mathcal{H}_\chi$ such that $M \subseteq N$. By the assumption $B \cap N$ is projective. Since $B \cap M \le_\sigma B \cap N$, it follows by Corollary 2.3.6 and Corollary 2.2.7, that $B \cap M$ is projective.

$\mathcal{3}) \Rightarrow \mathcal{1}$): Let χ be large enough and let

$$I = \{\, M \prec \mathcal{H}_\chi \,:\, B \in M, \, |M| = \aleph_1 \,\}$$

be with the partial ordering: $M \leq N \Leftrightarrow M \subseteq N$. For any $M \in I$. Let $B_M = B \cap M$. Clearly $(B_M)_{M \in I}$ is a filtration of B. By the assumption, B_M is projective for every $M \in I$. By Theorem A.1.4, b), $(I, \leq)$ also satisfies $\mathcal{2})_{\aleph_2}$ of the definition of $\aleph_2$-projective filtration. $\qquad\Box$ (Theorem A.4.7)

Corollary A.4.8 *If B is $\aleph_2$-projectively filtered and $A \leq_\sigma B$ then also A is $\aleph_2$-projectively filtered.*

Proof Let χ be large enough. Similarly to the proof of $\mathcal{3}) \Rightarrow \mathcal{1}$) of Theorem A.4.7, it is enough to show that for any $M \prec \mathcal{H}_\chi$, if $A, B \in M$ and $|M| = \aleph_1$ then $A \cap M$ is projective. Now let M be such elementary substructure of $\mathcal{H}_\chi$. Since B is $\aleph_2$-projectively filtered, $B \cap M$ is projective by Theorem A.4.7. By $M \models$ "$A \leq_\sigma B$" and by Lemma A.1.11, we have $A \cap M \leq_\sigma B \cap M$. Hence, by Corollary 2.3.6 and Corollary 2.2.7, $A \cap M$ is also projective. $\qquad\Box$ (Corollary A.4.8)

Now let us turn to the consistency result mentioned at the beginning of this section. The result uses the following axiom known as *Fleissner's Axiom R*. Let us say that $T \subseteq [X]^{\aleph_1}$ is *tight* if, for every increasing sequence $(u_\alpha)_{\alpha < \omega_1}$ of elements of T of length ω_1, $\bigcup_{\alpha < \omega_1} u_\alpha \in T$ holds.

(Axiom R):   *Suppose $\lambda \geq \aleph_2$, $\operatorname{cf}\lambda > \omega$ and $T \subseteq [\lambda]^{\aleph_1}$ is cofinal in $([\lambda]^{\aleph_1}, \subseteq)$ and tight. Then for any stationary $S \subseteq [\lambda]^{\aleph_0}$, there is $X \in T$ such that $S \cap [X]^{\aleph_0}$ is stationary in $[X]^{\aleph_0}$.*

As we shall see later, we need rather large cardinals to establish the consistency of Axiom R. In particular, its consistency cannot be proved merely from the assumption of the consistency of the usual axiom system of set theory (ZFC). Axiom R is a consequence of Martin's Maximum whose consistency can be proved under the assumption of the consistency of ZFC+ "$\exists$ supercompact cardinal". While $2^{\aleph_0} = \aleph_2$ follows from Martin's Maximum, the consistency of Axiom R + CH can be also shown under the assumption of the consistency of ZFC+ "$\exists$ supercompact cardinal". For more details see [1]. The following theorem was first proved under the stronger assumption $MA^+(\sigma\text{-closed})$ in [17]. Qi Feng then pointed out that the proof for Boolean algebras B with $\operatorname{cf}|B| > \omega$ (Case I in the proof below) can be done under Axiom R. The following complete proof appears here for the first time.

Theorem A.4.9 (Feng, Fuchino) (Axiom R) *A Boolean algebra B is $\aleph_2$-projectively filtered if and only if B is rc-filtered.*

Proof If B is rc-filtered then B is $\aleph_2$-projectively filtered by Lemma A.3.1, a).

Assume now that B is $\aleph_2$-projectively filtered. If $|B| \leq \aleph_1$ then any $\aleph_2$-projective filtration of B contains B itself hence B is rc-filtered. Hence we may assume $|B| \geq \aleph_2$. We consider first:

Case I: $\operatorname{cf}|B| > \omega$.

Let
$$T = \{\, A : A \leq_{\mathrm{rc}} B,\ A \text{ is projective and } |A| = \aleph_1 \,\}.$$

By Lemma A.2.6 and Theorem A.4.7, T is cofinal in $([B]^{\aleph_1}, \subseteq)$. T is tight: suppose that $(A_\alpha)_{\alpha<\omega_1}$ is an increasing sequence in T. Let $A = \bigcup_{\alpha<\omega_1} A_\alpha$. If A were not relatively complete in B then there would be $b \in B$ such that $(p^B_{A_\alpha}(b))_{\alpha<\omega_1}$ is a non eventually constant increasing sequence in A. But this is a contradiction since B satisfies the ccc. Hence we have $A \leq_{\mathrm{rc}} B$. Since T is cofinal in $[B]^{\aleph_1}$, there is some $A' \leq B$ such that $A \leq A'$ and A' is projective. Since $A \leq_{\mathrm{rc}} A'$ and $|A| = \aleph_1$ it follows that A is also projective. Hence $A \in T$.

Now, if B were not rc-filtered, then
$$S = \{\, C \leq B : |C| = \aleph_0 \text{ and } C \text{ is not relatively complete in } B \,\}$$

would be stationary in $[B]^{\aleph_0}$. Hence by Axiom R, there is $A \in T$ such that $S \cap [A]^{\aleph_0}$ is stationary in $[A]^{\aleph_0}$. But, since $A \leq_{\mathrm{rc}} B$, each $C \in S \cap [A]^{\aleph_0}$ is not relatively complete in A. This is a contradiction to the fact that A is projective.

Case II: $\operatorname{cf}|B| = \omega$.

Let $\kappa = |B|$ and let $\kappa = \sup_{n\in\omega} \kappa_n$ for an increasing sequence $(\kappa_n)_{n\in\omega}$ of uncountable regular cardinals. Since B is σ-filtered by Corollary A.4.6, there is a mapping $f : B \to [B]^{\leq\aleph_0}$ which witnesses (WFN) of B. Let $(B_n)_{n\in\omega}$ be an increasing sequence of subalgebras of B such that

1) $|B_n| = \kappa_n$ for every $n \in \omega$;

2) B_n is closed with respect to f for every $n \in \omega$ and

3) $B = \bigcup_{n\in\omega} B_n$.

By 2), and by similar argument to that for the Claim in the proof of Theorem 2.2.3, $B_n \leq_\sigma B$ holds for every $n \in \omega$. Hence by Corollary A.4.8, B_n is $\aleph_2$-projectively filtered for every $n \in \omega$. So by 1) and Case I, B_n, $n \in \omega$ are rc-filtered. Hence by Bandlow's Theorem (Theorem 2.2.11), it follows that B is also rc-filtered. $\square$ (Theorem A.4.9)

By Theorem A.4.9 and (by the proof of) Theorem A.3.2 we obtain:

Theorem A.4.10 (Axiom R) E^κ_ω *does not hold for any regular* $\kappa \geq \aleph_2$.

It is known that we need rather large cardinals to establish the conclusion of Theorem A.4.10 (for more see e.g. [33]). In particular, this shows that we need large cardinals to obtain the consistency of Axiom R.

We can extend the list of characterizations of $\aleph_2$-projectively filtered Boolean algebras by the following:

Lemma A.4.11 *For a Boolean algebra* B *the following are equivalent:*

1) B *is* $\aleph_2$-*projectively filtered;*

2) *for every countable* $C \leq B$ *there exists a countable* $C' \leq B$ *such that* $C \leq C' \leq_{\mathrm{rc}} B$, *and every relatively complete subalgebra* A *of* B *of cardinality* $\aleph_1$ *is projective.*

Proof $1) \Rightarrow 2$): Assume that B is $\aleph_2$-projectively filtered. If $C \leq B$ is countable then, by Theorem A.4.7, 2) and Lemma A.2.6, there exists $A \leq_{rc} B$ such that $C \leq A$ and A is projective. Take countable $C' \leq A$ such that $C \leq C'$ and $C' \leq_{rc} A$. Then we have $C' \leq_{rc} B$. If $A \leq_{rc} B$ is of cardinality $\aleph_1$, again by Theorem A.4.7, 2) and Lemma A.2.6, we can find $A' \leq_{rc} B$ such that $A \leq A'$ and A' is projective. Since we have $A \leq_{rc} A'$ and $|A| = \aleph_1$, it follows that A is also projective.

$2) \Rightarrow 1$): Assume that 2) holds for B. Let χ be large enough. Let $M \prec \mathcal{H}_\chi$ be V_{ω_1}-like over B and let $(M_\alpha)_{\alpha < \omega_1}$ be a witness of V_{ω_1}-likeness of M over B. Since $B \cap M_\alpha \in M_{\alpha+1}$ and $M_{\alpha+1} \models$ "$B \cap M_\alpha$ is countable", by the assumption there exists a countable $C_\alpha \in M_{\alpha+1}$ such that $M_{\alpha+1} \models$ "$B \cap M_\alpha \leq C_\alpha \leq_{rc} B$" for every $\alpha < \omega_1$. By Lemma A.1.11, it follows that

$$(+) \quad B \cap M_\alpha \leq C_\alpha \leq B \cap M_{\alpha+1}.$$

By $M_{\alpha+1} \prec \mathcal{H}_\chi$ (and absoluteness of "$x \leq_{rc} y$") we have $C_\alpha \leq_{rc} B$. By $(+)$, $(C_\alpha)_{\alpha < \omega_1}$ is an increasing sequence of relatively complete subalgebras of B and $B \cap M = \bigcup_{\alpha < \omega_1} C_\alpha$ holds. Thus we have $B \cap M \leq_{rc} B$. By 2), it follows that $B \cap M$ is projective. Hence by Theorem A.4.7, B is $\aleph_2$-projectively filtered.
                                                                    $\square$ (Lemma A.4.11)

The characterization of $\aleph_2$-projectively filtered Boolean algebras above has the following application: For a cardinal κ let $L_{\infty\kappa}$ be the logic whose formulas are constructed recursively just as in first-order logic, but with the difference that conjunction and disjunction of any set of formulas as well as quantification over any block of free variables of cardinality $< \kappa$ is allowed. The relation '$\models$' for $L_{\infty\kappa}$ is defined in a canonical way similarly to the first-order case. A Boolean algebra B is called $L_{\infty\kappa}$-*projective* if B is $L_{\infty\kappa}$-elementary equivalent to a projective Boolean algebra, that is, for some projective Boolean algebra A we have $B \models \varphi \Leftrightarrow A \models \varphi$ for every $L_{\infty\kappa}$ sentence φ in the vocabulary of Boolean algebras. Similarly we say that a Boolean algebra B is $L_{\infty\kappa}$-*free* if B is $L_{\infty\kappa}$-elementary equivalent to $\mathrm{Fr}\,\kappa$. It can be shown easily that if B is $L_{\infty\kappa}$-projective then $B \otimes \mathrm{Fr}\,\kappa$ is $L_{\infty\kappa}$-free. Being $L_{\infty\kappa}$-projective or $L_{\infty\kappa}$-free can also be seen as a kind of "almost projectivity". For more about $L_{\infty\kappa}$-projectiveness and $L_{\infty\kappa}$-freeness, see [18] and [20]. It is easily seen that the assertion in Lemma A.4.11, 2) can be formulated in an $L_{\infty\aleph_2}$-sentence and this is true in every projective Boolean algebra by Lemma A.3.1, b). Hence we obtain:

Proposition A.4.12 *Every $L_{\infty\aleph_2}$-projective Boolean algebra is $\aleph_2$-projectively filtered.* $\square$

From Proposition A.4.12 and Theorem A.4.9, it follows that:

Corollary A.4.13 (Axiom R) *Every $L_{\infty\aleph_2}$-projective Boolean algebra is rc-filtered.* $\square$

On the other hand, the Boolean algebras constructed in Theorem A.3.2 are $L_{\infty\kappa}$-free (this follows from an algebraic characterization of $L_{\infty\kappa}$-freeness, see [18]). Hence:

Theorem A.4.14 $(V = L)$ *For every* κ, *there is* $L_{\infty\kappa}$-*free Boolean algebra which is not rc-filtered.*   □

For Question 8, we cannot have any consistency result corresponding to Theorem A.4.9:

Theorem A.4.15 (Shelah) (in ZFC without any additional assumption) *There exists a non-projective Boolean algebra of size* $\aleph_3$ *which is* $\aleph_3$-*projectively filtered.*
□

Hence we obtain the negative answer to Question 8 already in ZFC. A proof of Theorem A.4.15 will be included in [20].

A.5   Forcing projectivity

In this section, we study the class of rc-filtered Boolean algebras in the light of the method of forcing. Compared with the preceding sections, this one is going to be less self-contained: we assume some elementary facts about forcing. Chapters VI, VII of [39] should cover what we need here.

A partially ordered set is called proper if, for any stationary $S \subseteq [X]^{\aleph_0}$ for some X, $\Vdash_P$ "S is stationary in $[X]^{\aleph_0}$" holds. It is easily seen that a proper partially ordered set preserves $\aleph_1$. ccc partially ordered sets and σ-closed partially ordered sets are examples of proper partially ordered sets (for more about properness the reader may consult [31]).

Theorem A.5.1 *For a Boolean algebra B, the following are equivalent:*

   a) B is rc-filtered;

   b)   There exists a proper partially ordered set P such that $\Vdash_P$ "B is projective";

   c)   If P is a proper partially ordered set collapsing $|B|$ to be $\leq \aleph_1$, then $\Vdash_P$ "B is projective".

Proof $a) \Rightarrow c$): Let B be rc-filtered and let $L, U : B \to [B]^{<\aleph_0}$ be witnesses of (FN) of B. Clearly we have $\Vdash_P$ "L and U witness (FN) of B". Hence we have $\Vdash_P$ "B is rc-filtered". Since $\Vdash_P$ "$|B| \leq \aleph_1$", it follows from Corollary 2.2.7 that $\Vdash_P$ "B is projective".

$c) \Rightarrow b$) is clear.

$b) \Rightarrow a$): Let P be a proper partially ordered set such that $\Vdash_P$ "B is projective". If B were not rc-filtered, then by Theorem 2.2.3

$$S' = \{ C \in [B]^{\aleph_0} : C \text{ is not a relatively complete subalgebra of } B \}$$

would be stationary in $[B]^{\aleph_0}$. Since $\{ C \leq B : |C| = \aleph_0 \}$ is closed unbounded in $[B]^{\aleph_0}$ it follows that

$$S = \{ C \leq B : |C| = \aleph_0, C \text{ is not relatively complete in } B \}$$

is also stationary in $[B]^{\aleph_0}$. Hence, by properness of P, it follows that $\Vdash_P$ "$\mathcal{S}$ is stationary in $[B]^{\aleph_0}$". Since we have $\Vdash_P$ "every $C \in \mathcal{S}$ is not relatively complete in B", we obtain $\Vdash_P$ "B is not rc-filtered". In particular, $\Vdash_P$ "B is not projective". This is a contradiction to our assumption. $\square$ (Theorem A.5.1)

As an application of Theorem A.5.1, let us consider the following theorem:

Theorem A.5.2 ([20]) *If B is a subalgebra of a rc-filtered Boolean algebra then either*

$$\mathcal{C} = \{\, C \le B \,:\, |C| = \aleph_0, \; C \text{ is relatively complete in } B \,\}$$

contains a closed unbounded subset of $[B]^{\aleph_0}$ or is disjoint to a closed unbounded subset of $[B]^{\aleph_0}$.

The Boolean algebras constructed in Proposition 2.10.1 and Theorem A.3.2 does not have the property in Theorem A.5.2. Hence, by Theorem A.5.2, they cannot be embedded into a rc-filtered Boolean algebra.

For the proof of Theorem A.5.2, we shall assume the following result by S. Shelah from [20]:

Theorem A.5.3 ([20]) *If B is a subalgebra of a projective Boolean algebra then either*

$$\mathcal{C} = \{\, C \le B \,:\, |C| = \aleph_0, \; C \text{ is relatively complete in } B \,\}$$

contains a closed unbounded subset of $[B]^{\aleph_0}$ or is disjoint to a closed unbounded subset of $[B]^{\aleph_0}$. $\square$

Proof of Theorem A.5.2 from Theorem A.5.3: Assume that B is a subalgebra of a rc-filtered Boolean algebra A. Further assume that $\mathcal{C} = \{\, C \le B \,:\, |C| = \aleph_0, \; C \text{ is relatively complete in } B \,\}$ does not contain any closed unbounded subset of $[B]^{\aleph_0}$. Let P be a σ-closed partially ordered set such that $\Vdash_P$ "$|A| \le \aleph_1$". Then $\Vdash_P$ "A is a projective Boolean algebra" by Theorem A.5.1. By the assumption and properness of P, we have $\Vdash_P$ "$[B]^{\aleph_0} \setminus \mathcal{C}$ is stationary". Hence by Theorem A.5.3, it follows that

$$\Vdash_P \text{ ``}\mathcal{C} \text{ is disjoint to a closed unbounded set of } [B]^{\aleph_0}\text{''}.$$

Let χ be large enough for B such that P, $P(P)$, $P(P(P))$, $\Vdash_P \in \mathcal{H}_\chi$.

By Theorem A.1.7, it follows from the next claim that $\mathcal{C}$ is disjoint to a closed unbounded subset of $[B]^{\aleph_0}$.

CLAIM 1. *For any countable $M \prec \mathcal{H}_\chi$ such that B, P, $\Vdash_P \in \mathcal{H}_\chi$, we have that $B \cap M$ is not relatively complete in B.*

$\vdash$ By Theorem A.1.16, we have $B \cap M \le B$. Let $B \cap M = \{\, b_n \,:\, n \in \omega \,\}$. Take a P-name $\dot{\mathcal{C}} \in M$ such that

$$\Vdash_P \text{ ``}\dot{\mathcal{C}} \text{ is a closed unbounded subset of } [B]^{\aleph_0} \text{ and } \forall C \in \dot{\mathcal{C}}(C \le_{\neg rc} B)\text{''}.$$

Let $f : \omega \to \omega^2$, $n \mapsto (f_0(n), f_1(n))$ be a surjection such that $f_0(n) \le n$ holds for every $n \in \omega$. We can construct sequences $(p_n)_{n \in \omega}$, $(\dot{B}_n)_{n \in \omega}$ and $(b_{n,m})_{n,m \in \omega}$

inductively such that

1) $(p_n)_{n\in\omega}$ is a decreasing sequence in P and $p_n \in M$ for every $n \in \omega$;

2) $(\dot{B}_n)_{n\in\omega}$ and $(\dot{b}_{n,m})_{n,m\in\omega}$ are sequence of P-names and $\dot{B}_n$, $\dot{b}_{n,m} \in M$ for every $n, m \in \omega$;

3) $p_n \Vdash_P$ " $\dot{B}_n \in \dot{C}$, $\dot{B}_n = \{\, \dot{b}_{n,m} : m \in \omega \,\}$ " for all $n \in \omega$;

4) $p_{n+1} \Vdash_P$ " $\dot{B}_n \leq \dot{B}_{n+1}$, $b_n \in \dot{B}_{n+1}$ " for every $n \in \omega$;

5) p_{n+1} decides $\dot{b}_{f_0(n),f_1(n)}$

(*i.e.* there is $b \in B$ such that $p_{n+1} \Vdash_P$ " $\dot{b}_{f_0(n),f_1(n)} = b$ ". Note that, we have $b \in M$ since p_{n+1}, $\Vdash_P$, $\dot{b}_{f_0(n),f_1(n)} \in M$).

As P is σ-closed, there is $p \in P$ such that $p \leq p_n$ for every $n \in \omega$. By 5), p decides every $\dot{b}_{n,m}$, $n, m \in \omega$. Hence, for every $n \in \omega$, there is a countable $C_n \leq B$ such that $p \Vdash_P$ " $\dot{B}_n = C_n$ " and $C_n \leq B \cap M$ by the remark after 5). Let $C = \bigcup_{n\in\omega} C_n$. By 4), we have $b_n \in C_{n+1}$ for every $n \in \omega$. Hence $C = B \cap M$. By 3), we have $p \Vdash_P$ " $C_n \in \dot{C}$ " for every $n \in \omega$. Hence $p \Vdash_P$ " $C \in \dot{C}$ ". It follows that $p \Vdash_P$ " $C \leq_{\neg\mathrm{rc}} B$ ". Thus we have $C \leq_{\neg\mathrm{rc}} B$. $\dashv$ (CLAIM 1.)

□ (Theorem A.5.2)

For a property $\mathcal{P}$ of partially ordered set and a property $\mathcal{Q}$ of Boolean algebra, let us say that a Boolean algebra B is "$\mathcal{P}$-potentially $\mathcal{Q}$" if there is a partially ordered set P with the property $\mathcal{P}$ such that $\Vdash_P$ " B has the property $\mathcal{Q}$ ". Thus, by Theorem A.5.1, we have: B is proper-potentially projective iff B is σ-closed potentially projective iff B is rc-filtered. One might hope that ccc-potentially projcetiveness would provide a new class of Boolean algebras. This is not the case:

Theorem A.5.4 ([15]) *A Boolean algebra B is ccc-potentially projective if and only if B is projective.*   □

Similarly we have:

Proposition A.5.5 *A Boolean algebra B is proper-potentially rc-filtered if and only if B is rc-filtered.*

Proof The proof of $a) \Rightarrow c$) of Theorem A.5.1 shows that, if B is rc-filtered, then $\Vdash_P$ " B is rc-filtered " holds for every partially ordered set P.

Conversely assume that $\Vdash_P$ " B is rc-filtered " for some proper partially ordered set P. Let $\dot{Q}$ be a P-name such that $\Vdash_P$ " $\dot{Q}$ is a proper partially ordered set collapsing $|B|$ to $\leq \aleph_1$ ". Then, by Theorem A.5.1, $\Vdash_{P*\dot{Q}}$ " B is projective ". Since $P * \dot{Q}$ is proper, it follows by Theorem A.5.1, that B is rc-filtered.

□ (Proposition A.5.5)

However, the corresponding theorem does not hold for the property "a subalgebra of a free Boolean algebra". Let us say that a Boolean algebra is *dyadic* if B is embeddable into a free Boolean algebra.

Theorem A.5.6 ([20]) *There is a ccc-potentially dyadic Boolean algebra B which is not dyadic.*   □

Bibliography

[1] R. E. Beaudoin, *Strong analogues of Martin's axiom imply Axiom R.* Journal of Symbolic Logic, 52:1(1987) 216-218.

[2] I. Bandlov[9], *Factorization theorems and 'strong' sequences in bicompacta.* Soviet Math. Doklady, 22:1(1980), 196-200
Original: Doklady Akad. Nauk SSSR, 253:5(1980), 1036-1040.

[3] I. Bandlow, *On mappings that are co-absolute to projections along the Cantor perfect set.* (in Russian), Vestn. Mosk. Univ., Mat. Series, 1989:1, 40-44.

[4] I. Bandlow, *Anwendungen elementarer Unterstrukturen in der mengentheoretischen Topologie.* Habilitationsschrift, Universität Greifswald 1990.

[5] I. Bandlow, *A construction in set-theoretic topology by means of elementary substructures.* Zeitschr. Math. Logik, 37(1991), 467-480 .

[6] M.G. Bell, *Special points in compact spaces.* submitted 1993.

[7] C: C. Chang and H. J. Keisler, *Model theory.* North-Holland, Amsterdam etc. 1973.

[8] K. J. Devlin, *Aspects of constructibility.* LNM 354, Springer, Berlin etc. 1973.

[9] A. Dow, *An introduction to applications of elementary submodels to topology.* Topology Proceedings, 13(1988), 17-72.

[10] A. N. Dranišnikov, *Absolute extensors in dimension n and dimension raising n-soft maps.* Russian mathematical surveys, 39:5(1984), 63-111,
Original: Uspehi Mat. Nauk, 39:5(1984), 55-95.

[11] P. C. Eklof and A. H. Mekler *Almost free modules.* North-Holland, Amsterdam etc. 1990.

[12] R. Engelking, *Cartesian products and dyadic spaces.* Fundamenta Mathematicae 57(1965), 287-303.

---

[9]The 'v' is a mistake of the AMS translator; it is the same Ingo Bandlow as below

[13] R. Engelking, *General Topoplogy*. PWN, Warsaw 1977.

[14] R. Freese and J. B. Nation, *Projective lattices*.
Pacific Journal of Mathematics, 75(1978), 93-106.

[15] S. Fuchino *On potential embedding and versions of Martin's axiom*.
Notre Dame Journal of Logic, 33:4(1992), 481-492.

[16] S. Fuchino *Some problems of Ščepin on openly generated Boolean algebras*.
in: M. Weese and H. Wolter (eds.) *Proceedings of the Tenth Easter Conference on Model Theory*, Seminarberichte 93-1, Fachbereich Mathematik der Humboldt-Universität zu Berlin (1993), 14-29.

[17] S. Fuchino, *Some remarks on openly generated Boolean algebras*.
Journal of Symbolic Logic, 59(1994), 302-310.

[18] S. Fuchino, S. Koppelberg, and M. Takahashi *On $L_{\infty\kappa}$-free Boolean algebras*. Annals of Pure and Applied Logic, 55(1992), 265-284.

[19] S. Fuchino, S. Koppelberg, S. Shelah, *Boolean algebras with the weak Freese-Nation property*. preprint 1994.

[20] S. Fuchino, S. Shelah, *More on $L_{\infty\kappa}$-free Boolean algebras*. in preparation.

[21] S. Fuchino, S. Shelah, *The Number of openly generated Boolean algebras*. in preparation.

[22] P. R. Halmos, *Injective and projective Boolean algebras*.
Proc. Symp. Pure Math., 2(1961), 114-122.

[23] R. Haydon, *On a problem of Pełczynski: Miljutin spaces, Dugundji spaces and AE(0-dim)*. Studia Mathematica, 52(1974), 23-31.

[24] L. Heindorf, *Boolean semigroup rings and exponentials of compact zero-dimensional spaces*. Fundamenta Mathcmaticac, 153(1990), 37-47.

[25] L. Heindorf, *Zero-dimensional Dugundji spaces admit profinite lattice structures*. Comm. Math. Univ. Carolinae, 33(1992), 329-334.

[26] A. V. Ivanov, *On bicompacta all finite powers of which are hereditarily separable*. Soviet Math. Doklady, 19:6(1978), 1470-1473,
Original: Doklady Akad. Nauk SSSR, 243(1978), 1109-1112.

[27] A. V. Ivanov, *Superextensions of openly generated compact Hausdorff spaces*. Soviet Math. Doklady, 24:1(1981), 60-64,
Original: Doklady Akad. Nauk SSSR, 259(1981), 275-278.

[28] A. V. Ivanov, *Superextension-type functors and soft mappings*.
Mathematical Notes, 36(1984), 543-546,
Original: Mat. Zametki, 36(1984), 103-107.

[29] T. Iwamura, *A lemma on directed sets.* (in Japanese)
Zenkoku Shijo Sugaku Danwakai, 262(1944), 107-111

[30] T. Jech, *Set Theory.* Academic Press, New York etc. 1978.

[31] T. Jech, *Multiple Forcing.* Cambridge University Press 1986.

[32] R. B. Jensen, *The fine structure of the constructible hierarchy.*
Annals of Mathematical Logic, 4(1972), 229-308.

[33] A. Kanamori, *The Higher Infinite.* Springer, to appear 1994.

[34] S. Koppelberg, *General theory of Boolean algebras.*
vol. 1 of: J. D. Monk with R. Bonnet (Eds.), *Handbook of Boolean algebras,*
North-Holland, Amsterdam etc. 1989.

[35] S. Koppelberg, *Projective Boolean algebras.*
Chapter 20, 741-774, in vol. 3 of: J. D. Monk with R. Bonnet (Eds.),
Handbook of Boolean algebras, North-Holland, Amsterdam etc. 1989.

[36] S. Koppelberg, *Characterizations of Cohen Algebras.*
in: S. Andima et al. (Eds.), *Papers on General Topology and Applications*
Annals of the New York Academy of Sciences, 704(1993), 222-237.

[37] S. Koppelberg and S. Shelah *Subalgebras of Cohen algebras need not be Cohen.*
To appear in the Proceedings of the ASL-Logic Colloquium, Keele 1993.

[38] D. W. Kueker, *Countable approximations and Löwenheim-Skolem Theorems.* Annales of Mathematical Logic, 11(1977), 57-103.

[39] K. Kunen, *Set Theory.* North-Holland, Amsterdam etc. 1980.

[40] K. Kuratowski, *Topoplogy* vol. I, PWN, Warsaw 1966.

[41] E. V. Moiseev, *On spaces of closed growth and inclusion hyperspaces.*
Moscow University Math. Bull., 43(1988), 48-51,
Original: Vestn. Mosk. Univ., Mat. Series, 1988:3, 54-57.

[42] J. D. Monk, *Cardinal functions on Boolean algebras.*
Birkhäuser, Basel 1990.

[43] V.V. Pašenkov, *Extensions of compact spaces.*
Soviet Math. Doklady, 15:1(1974), 43-47
Original: Doklady Akad. Nauk 214:1(1974), 44-47.

[44] A. Pełczynski, *Linear extensions, linear avaragings and their applications to linear topological classification of spaces of continuous functions.*
Diss. Math., 58(1968), 1-90.

[45] I. V. Ponomarev, *On spaces co-absolute with metric spaces.*
Russian mathematical surveys, 21(1966), 87-114,
Original: Uspehi Mat. Nauk, 11:4(1966), 101-132.

[46] L. B . Shapiro, *A counter-example in the theory of dyadic bicompacta*
Russian mathematical surveys, 40(1985), 239-240,
Original: Uspehi Mat. Nauk, 40:5(1985), 267-268.

[47] L. B. Šapiro, *The space of closed subsets of $D^{\aleph_2}$ is not a dyadic bicompact.*
Soviet Math. Doklady, 17(1976), 937-941,
Original: Doklady Akad. Nauk SSSR, 228(1976), 1302-1305.

[48] L. B. Šapiro, *On spaces of closed subsets of bicompacts.*
Soviet Math. Doklady, 17(1976), 1567-1570,
Original: Doklady Akad. Nauk SSSR, 231:2(1976), 295-298.

[49] L. B. Shapiro, *On spaces coabsolute to dyadic compacta.*
Soviet Math. Doklady, 35(1987), 434-438,
Original: Doklady Akad. Nauk SSSR, 293:5(1987), 1077-1081.

[50] L. B. Shapiro, *On the homogeneity of hyperspaces of dyadic bicompacta.*
Mathematical Notes 49:1(1991), 85-88,
Original: Mat. Zametki, 49:1(1991), 120-126.

[51] L. B. Shapiro, *On the homogeneity of dyadic compact Hausdorff spaces.*
Mathematical Notes 54:4(1993), 1058-1072,
Original: Mat. Zametki, 54:4(1993), 117-139.

[52] L. B. Shapiro, *On κ-metrizable compact Hausdorff spaces*
Russ. Math. Doklady, to appear in vol. 48(1994),
Original: Doklady Russ. Akad. Nauk , 333:3(1993), 308-311.

[53] E. V. Ščepin, *Toplogy of limit spaces of uncountable inverse spectra.*
Russian mathematical surveys, 31(1976), 155-191,
Original: Uspehi Mat. Nauk, 31:5(1981), 191-226.

[54] E. V. Ščepin, *On κ-metrizable spaces.*
Mathematics of the USSR Izvestija 14(1980), 407-440,
Original: Izv. Akad. Nauk Ser. Mat., 43(2)(1979), 442-478.

[55] E. V. Ščepin, *Functors and uncountable powers of compacts.*
Russian mathematical surveys, 36(1981), 1-71,
Original: Uspehi Mat. Nauk, 36(3)(1981), 3-62.

[56] L. V. Širokov, *An extrinsic characterization of Dugundji spaces and κ-metrizable compact Hausdorff spaces.*
Soviet Math. Doklady, 25(1982), 507-510,
Original: Doklady Akad. Nauk SSSR, 263(1982), 1073-1077.

[57] S. Sirota, *Spectral representation of spaces of closed subsets of bicompacta.* Soviet Math. Doklady, 9(1968), 997-1000
Original: Doklady Akad. Nauk SSSR, 181:5(1968), 1069-1072.

[58] R. Solovay and S. Tennenbaum *Iterated Cohen extensions and Souslin's problem.* Annals of Mathematics, 94(1971), 201-245.

Index

Special Symbols

Subject index

Lecture Notes in Mathematics

For information about Vols. 1–1414
please contact your bookseller or Springer-Verlag

Vol. 1454: F. Baldassari, S. Bosch, B. Dwork (Eds.), p-adic Analysis. Proceedings, 1989. V, 382 pages. 1990.

Vol. 1455: J.-P. Françoise, R. Roussarie (Eds.), Bifurcations of Planar Vector Fields. Proceedings, 1989. VI, 396 pages. 1990.

Vol. 1456: L.G. Kovács (Ed.), Groups – Canberra 1989. Proceedings. XII, 198 pages. 1990.

Vol. 1457: O. Axelsson, L.Yu. Kolotilina (Eds.), Preconditioned Conjugate Gradient Methods. Proceedings, 1989. V, 196 pages. 1990.

Vol. 1458: R. Schaaf, Global Solution Branches of Two Point Boundary Value Problems. XIX, 141 pages. 1990.

Vol. 1459: D. Tiba, Optimal Control of Nonsmooth Distributed Parameter Systems. VII, 159 pages. 1990.

Vol. 1460: G. Toscani, V. Boffi, S. Rionero (Eds.), Mathematical Aspects of Fluid Plasma Dynamics. Proceedings, 1988. V, 221 pages. 1991.

Vol. 1461: R. Gorenflo, S. Vessella, Abel Integral Equations. VII, 215 pages. 1991.

Vol. 1462: D. Mond, J. Montaldi (Eds.), Singularity Theory and its Applications. Warwick 1989, Part I. VIII, 405 pages. 1991.

Vol. 1463: R. Roberts, I. Stewart (Eds.), Singularity Theory and its Applications. Warwick 1989, Part II. VIII, 322 pages. 1991.

Vol. 1464: D. L. Burkholder, E. Pardoux, A. Sznitman, Ecole d'Eté de Probabilités de Saint- Flour XIX-1989. Editor: P. L. Hennequin. VI, 256 pages. 1991.

Vol. 1465: G. David, Wavelets and Singular Integrals on Curves and Surfaces. X, 107 pages. 1991.

Vol. 1466: W. Banaszczyk, Additive Subgroups of Topological Vector Spaces. VII, 178 pages. 1991.

Vol. 1467: W. M. Schmidt, Diophantine Approximations and Diophantine Equations. VIII, 217 pages. 1991.

Vol. 1468: J. Noguchi, T. Ohsawa (Eds.), Prospects in Complex Geometry. Proceedings, 1989. VII, 421 pages. 1991.

Vol. 1469: J. Lindenstrauss, V. D. Milman (Eds.), Geometric Aspects of Functional Analysis. Seminar 1989-90. XI, 191 pages. 1991.

Vol. 1470: E. Odell, H. Rosenthal (Eds.), Functional Analysis. Proceedings, 1987-89. VII, 199 pages. 1991.

Vol. 1471: A. A. Panchishkin, Non-Archimedean L-Functions of Siegel and Hilbert Modular Forms. VII, 157 pages. 1991.

Vol. 1472: T. T. Nielsen, Bose Algebras: The Complex and Real Wave Representations. V, 132 pages. 1991.

Vol. 1473: Y. Hino, S. Murakami, T. Naito, Functional Differential Equations with Infinite Delay. X, 317 pages. 1991.

Vol. 1474: S. Jackowski, B. Oliver, K. Pawałowski (Eds.), Algebraic Topology, Poznań 1989. Proceedings. VIII, 397 pages. 1991.

Vol. 1475: S. Busenberg, M. Martelli (Eds.), Delay Differential Equations and Dynamical Systems. Proceedings, 1990. VIII, 249 pages. 1991.

Vol. 1476: M. Bekkali, Topics in Set Theory. VII, 120 pages. 1991.

Vol. 1477: R. Jajte, Strong Limit Theorems in Noncommutative L_2-Spaces. X, 113 pages. 1991.

Vol. 1478: M.-P. Malliavin (Ed.), Topics in Invariant Theory. Seminar 1989-1990. VI, 272 pages. 1991.

Vol. 1479: S. Bloch, I. Dolgachev, W. Fulton (Eds.), Algebraic Geometry. Proceedings, 1989. VII, 300 pages. 1991.

Vol. 1480: F. Dumortier, R. Roussarie, J. Sotomayor, H. Żoładek, Bifurcations of Planar Vector Fields: Nilpotent Singularities and Abelian Integrals. VIII, 226 pages. 1991.

Vol. 1481: D. Ferus, U. Pinkall, U. Simon, B. Wegner (Eds.), Global Differential Geometry and Global Analysis. Proceedings, 1991. VIII, 283 pages. 1991.

Vol. 1482: J. Chabrowski, The Dirichlet Problem with L^2-Boundary Data for Elliptic Linear Equations. VI, 173 pages. 1991.

Vol. 1483: E. Reithmeier, Periodic Solutions of Nonlinear Dynamical Systems. VI, 171 pages. 1991.

Vol. 1484: H. Delfs, Homology of Locally Semialgebraic Spaces. IX, 136 pages. 1991.

Vol. 1485: J. Azéma, P. A. Meyer, M. Yor (Eds.), Séminaire de Probabilités XXV. VIII, 440 pages. 1991.

Vol. 1486: L. Arnold, H. Crauel, J.-P. Eckmann (Eds.), Lyapunov Exponents. Proceedings, 1990. VIII, 365 pages. 1991.

Vol. 1487: E. Freitag, Singular Modular Forms and Theta Relations. VI, 172 pages. 1991.

Vol. 1488: A. Carboni, M. C. Pedicchio, G. Rosolini (Eds.), Category Theory. Proceedings, 1990. VII, 494 pages. 1991.

Vol. 1489: A. Mielke, Hamiltonian and Lagrangian Flows on Center Manifolds. X, 140 pages. 1991.

Vol. 1490: K. Metsch, Linear Spaces with Few Lines. XIII, 196 pages. 1991.

Vol. 1491: E. Lluis-Puebla, J.-L. Loday, H. Gillet, C. Soulé, V. Snaith, Higher Algebraic K-Theory: an overview. IX, 164 pages. 1992.

Vol. 1492: K. R. Wicks, Fractals and Hyperspaces. VIII, 168 pages. 1991.

Vol. 1493: E. Benoît (Ed.), Dynamic Bifurcations. Proceedings, Luminy 1990. VII, 219 pages. 1991.

Vol. 1494: M.-T. Cheng, X.-W. Zhou, D.-G. Deng (Eds.), Harmonic Analysis. Proceedings, 1988. IX, 226 pages. 1991.

Vol. 1495: J. M. Bony, G. Grubb, L. Hörmander, H. Komatsu, J. Sjöstrand, Microlocal Analysis and Applications. Montecatini Terme, 1989. Editors: L. Cattabriga, L. Rodino. VII, 349 pages. 1991.

Vol. 1496: C. Foias, B. Francis, J. W. Helton, H. Kwakernaak, J. B. Pearson, H_∞-Control Theory. Como, 1990. Editors: E. Mosca, L. Pandolfi. VII, 336 pages. 1991.

Vol. 1497: G. T. Herman, A. K. Louis, F. Natterer (Eds.), Mathematical Methods in Tomography. Proceedings 1990. X, 268 pages. 1991.

Vol. 1498: R. Lang, Spectral Theory of Random Schrödinger Operators. X, 125 pages. 1991.

Vol. 1499: K. Taira, Boundary Value Problems and Markov Processes. IX, 132 pages. 1991.

Vol. 1500: J.-P. Serre, Lie Algebras and Lie Groups. VII, 168 pages. 1992.

Vol. 1501: A. De Masi, E. Presutti, Mathematical Methods for Hydrodynamic Limits. IX, 196 pages. 1991.

Vol. 1502: C. Simpson, Asymptotic Behavior of Mono-dromy. V, 139 pages. 1991.

Vol. 1503: S. Shokranian, The Selberg-Arthur Trace Formula (Lectures by J. Arthur). VII, 97 pages. 1991.

Vol. 1504: J. Cheeger, M. Gromov, C. Okonek, P. Pansu, Geometric Topology: Recent Developments. Editors: P. de Bartolomeis, F. Tricerri. VII, 197 pages. 1991.

Vol. 1505: K. Kajitani, T. Nishitani, The Hyperbolic Cauchy Problem. VII, 168 pages. 1991.

Vol. 1506: A. Buium, Differential Algebraic Groups of Finite Dimension. XV, 145 pages. 1992.

Vol. 1507: K. Hulek, T. Peternell, M. Schneider, F.-O. Schreyer (Eds.), Complex Algebraic Varieties. Proceedings, 1990. VII, 179 pages. 1992.

Vol. 1508: M. Vuorinen (Ed.), Quasiconformal Space Mappings. A Collection of Surveys 1960-1990. IX, 148 pages. 1992.

Vol. 1509: J. Aguadé, M. Castellet, F. R. Cohen (Eds.), Algebraic Topology - Homotopy and Group Cohomology. Proceedings, 1990. X, 330 pages. 1992.

Vol. 1510: P. P. Kulish (Ed.), Quantum Groups. Proceedings, 1990. XII, 398 pages. 1992.

Vol. 1511: B. S. Yadav, D. Singh (Eds.), Functional Analysis and Operator Theory. Proceedings, 1990. VIII, 223 pages. 1992.

Vol. 1512: L. M. Adleman, M.-D. A. Huang, Primality Testing and Abelian Varieties Over Finite Fields. VII, 142 pages. 1992.

Vol. 1513: L. S. Block, W. A. Coppel, Dynamics in One Dimension. VIII, 249 pages. 1992.

Vol. 1514: U. Krengel, K. Richter, V. Warstat (Eds.), Ergodic Theory and Related Topics III, Proceedings, 1990. VIII, 236 pages. 1992.

Vol. 1515: E. Ballico, F. Catanese, C. Ciliberto (Eds.), Classification of Irregular Varieties. Proceedings, 1990. VII, 149 pages. 1992.

Vol. 1516: R. A. Lorentz, Multivariate Birkhoff Interpolation. IX, 192 pages. 1992.

Vol. 1517: K. Keimel, W. Roth, Ordered Cones and Approximation. VI, 134 pages. 1992.

Vol. 1518: H. Stichtenoth, M. A. Tsfasman (Eds.), Coding Theory and Algebraic Geometry. Proceedings, 1991. VIII, 223 pages. 1992.

Vol. 1519: M. W. Short, The Primitive Soluble Permutation Groups of Degree less than 256. IX, 145 pages. 1992.

Vol. 1520: Yu. G. Borisovich, Yu. E. Gliklikh (Eds.), Global Analysis – Studies and Applications V. VII, 284 pages. 1992.

Vol. 1521: S. Busenberg, B. Forte, H. K. Kuiken, Mathematical Modelling of Industrial Process. Bari, 1990. Editors: V. Capasso, A. Fasano. VII, 162 pages. 1992.

Vol. 1522: J.-M. Delort, F. B. I. Transformation. VII, 101 pages. 1992.

Vol. 1523: W. Xue, Rings with Morita Duality. X, 168 pages. 1992.

Vol. 1524: M. Coste, L. Mahé, M.-F. Roy (Eds.), Real Algebraic Geometry. Proceedings, 1991. VIII, 418 pages. 1992.

Vol. 1525: C. Casacuberta, M. Castellet (Eds.), Mathematical Research Today and Tomorrow. VII, 112 pages. 1992.

Vol. 1526: J. Azéma, P. A. Meyer, M. Yor (Eds.), Séminaire de Probabilités XXVI. X, 633 pages. 1992.

Vol. 1527: M. I. Freidlin, J.-F. Le Gall, Ecole d'Eté de Probabilités de Saint-Flour XX – 1990. Editor: P. L. Hennequin. VIII, 244 pages. 1992.

Vol. 1528: G. Isac, Complementarity Problems. VI, 297 pages. 1992.

Vol. 1529: J. van Neerven, The Adjoint of a Semigroup of Linear Operators. X, 195 pages. 1992.

Vol. 1530: J. G. Heywood, K. Masuda, R. Rautmann, S. A. Solonnikov (Eds.), The Navier-Stokes Equations II – Theory and Numerical Methods. IX, 322 pages. 1992.

Vol. 1531: M. Stoer, Design of Survivable Networks. IV, 206 pages. 1992.

Vol. 1532: J. F. Colombeau, Multiplication of Distributions. X, 184 pages. 1992.

Vol. 1533: P. Jipsen, H. Rose, Varieties of Lattices. X, 162 pages. 1992.

Vol. 1534: C. Greither, Cyclic Galois Extensions of Commutative Rings. X, 145 pages. 1992.

Vol. 1535: A. B. Evans, Orthomorphism Graphs of Groups. VIII, 114 pages. 1992.

Vol. 1536: M. K. Kwong, A. Zettl, Norm Inequalities for Derivatives and Differences. VII, 150 pages. 1992.

Vol. 1537: P. Fitzpatrick, M. Martelli, J. Mawhin, R. Nussbaum, Topological Methods for Ordinary Differential Equations. Montecatini Terme, 1991. Editors: M. Furi, P. Zecca. VII, 218 pages. 1993.

Vol. 1538: P.-A. Meyer, Quantum Probability for Probabilists. X, 287 pages. 1993.

Vol. 1539: M. Coornaert, A. Papadopoulos, Symbolic Dynamics and Hyperbolic Groups. VIII, 138 pages. 1993.

Vol. 1540: H. Komatsu (Ed.), Functional Analysis and Related Topics, 1991. Proceedings. XXI, 413 pages. 1993.

Vol. 1541: D. A. Dawson, B. Maisonneuve, J. Spencer, Ecole d' Eté de Probabilités de Saint-Flour XXI - 1991. Editor: P. L. Hennequin. VIII, 356 pages. 1993.

Vol. 1542: J.Fröhlich, Th.Kerler, Quantum Groups, Quantum Categories and Quantum Field Theory. VII, 431 pages. 1993.

Vol. 1543: A. L. Dontchev, T. Zolezzi, Well-Posed Optimization Problems. XII, 421 pages. 1993.

Vol. 1544: M.Schürmann, White Noise on Bialgebras. VII, 146 pages. 1993.

Vol. 1545: J. Morgan, K. O'Grady, Differential Topology of Complex Surfaces. VIII, 224 pages. 1993.

Vol. 1546: V. V. Kalashnikov, V. M. Zolotarev (Eds.), Stability Problems for Stochastic Models. Proceedings, 1991. VIII, 229 pages. 1993.

Vol. 1547: P. Harmand, D. Werner, W. Werner, M-ideals in Banach Spaces and Banach Algebras. VIII, 387 pages. 1993.

Vol. 1548: T. Urabe, Dynkin Graphs and Quadrilateral Singularities. VI, 233 pages. 1993.

Vol. 1549: G. Vainikko, Multidimensional Weakly Singular Integral Equations. XI, 159 pages. 1993.

Vol. 1550: A. A. Gonchar, E. B. Saff (Eds.), Methods of Approximation Theory in Complex Analysis and Mathematical Physics IV, 222 pages, 1993.